ESSENTIAL MATLAB® AND OCTAVE

ESSENTIAL MATLAB® AND OCTAVE

Jesús Rogel-Salazar

CRC Press
Taylor & Francis Group
Boca Raton London New York

CRC Press is an imprint of the
Taylor & Francis Group, an **informa** business

MATLAB® is a trademark of The MathWorks, Inc. and is used with permission. The MathWorks does not warrant the accuracy of the text or exercises in this book. This book's use or discussion of MATLAB® software or related products does not constitute endorsement or sponsorship by The MathWorks of a particular pedagogical approach or particular use of the MATLAB® software.

MIX
Paper from
responsible sources
FSC® C014174

CRC Press
Taylor & Francis Group
6000 Broken Sound Parkway NW, Suite 300
Boca Raton, FL 33487-2742

© 2015 by Taylor & Francis Group, LLC
CRC Press is an imprint of Taylor & Francis Group, an Informa business

No claim to original U.S. Government works

Printed on acid-free paper
Version Date: 20140930

International Standard Book Number-13: 978-1-4822-3463-3 (Paperback)

To A. J. Johnson

Contents

List of Figures

List of Tables

Preface

THIS BOOK IS AN INTRODUCTION to the most essential aspects of MATLAB®[1] and GNU Octave. It is intended to be a companion to students in physics, mathematics, statistics, engineering and any other subjects that require the use of computers to solve numerical problems. The book addresses both MATLAB and Octave with the intention of making it easier for readers to implement their scripts and programmes in a way that is accessible to all. On the one hand MATLAB is a trademark of The MathWorks and as such is a proprietary software subject to licensing. On the other hand, GNU Octave shares many of the capabilities of MATLAB with the added merit of being freely distributed under the GNU General Public License.

The aim of the book is to provide straightforward explanations and examples that can be readily used by readers, and help them understand the elements of the software. The book therefore does not provide in-depth discussions of the implementation of specific algorithms or commands. The examples presented in this book have been tested in the latest versions of MATLAB (R2014a) and GNU Octave (3.8). It is important to clarify that although these two software packages share a large number of features, they are not strictly the same and therefore care must be taken when developing

[1] For product information please contact:

The MathWorks, Inc.
3 Apple Hill Drive
Natick, MA 01760-2098 USA
Tel: 508-647-7000
Fax: 508-647-7001
E-mail: info@mathworks.com
Web: www.mathworks.com

scripts in one of them with the intention of being used in the other. This is particularly true in the case of toolboxes available for MATLAB that may not have a counterpart in Octave. We assume that the reader is interacting with the computer by issuing commands on the form of successive lines of text (command lines). The code presented in this book has been written with the intention of being used in either of the packages. Throughout the text we present computer code enclosed in a box as such:

```
> 1 + 1 % Example of computer code

ans =
    2
```

We have made use of a diple (>) to denote the command line terminal prompt in either MATLAB or Octave, and the output is shown immediately below as it would appear in the command line of the software itself.

In cases where the code or output is specific to MATLAB or Octave we have added a note to that effect in the margin of the box surrounding the command or output. For MATLAB the box looks as follows:

```
> % A margin note for code specific to MATLAB
```

MATLAB

whereas for Octave the box is shown as:

```
> % A margin note for code specific to Octave
```

Octave

We have made use of margin notes, such as the one that appears to the right of this paragraph, to highlight certain topics or commands, as well as to provide some useful remarks. Please note that the output generated by MATLAB

This is an example of the margin notes used throughout this book.

and Octave may differ in formatting and for the purposes of the book we have decided to show a fixed number of significant figures that fit within the boxes.

THE BOOK ITSELF IS SET in a serial form where concepts introduced in earlier chapters are used in later ones. Nonetheless, the material is sufficiently self-contained to allow the reader to use the book as a reference tool. Having said that, the book is intended to be more than a technical manual and that is why we have framed examples and discussions with a scientific twist. Furthermore, given the nature of programming, it is impossible to cover every single intricacy and thus some points that are not explicitly addressed in the text are left to the reader as further practise in the exercises. It is important that the reader is aware that both Octave and particularly MATLAB have a number of toolboxes available to be used. We have not made use of these toolboxes in this book as the main emphasis is in general programming and scripting with both languages. Nonetheless we recommend the reader to take a look at the *Extra packages for GNU Octave* site for Octave and the *Mathworks* site for MATLAB.

Note that some points that are not explicitly covered in the main text are addressed in the Exercises sections at the end of each chapter.

Octave Packages available at
http://octave.sourceforge.net/packages.php

MATLAB Toolboxes available at
http://www.mathworks.co.uk/products/

We start Chapter 1 by introducing the software and outline its availability in various platforms. The chapter also serves as a foundation for the rest of the book presenting the basic structures and syntax of the software. The book follows, in Chapter 2, with a presentation of the simplest arrays that MATLAB and Octave deal with: vectors. We present the ways of manipulating the elements of a vector and show the most common operators on vectors. We then extend the discussion of the concept of a vector to that of a matrix in Chapter 3. Matrices are effectively the building blocks of the software itself. We introduce some special matrices and explain the operations that can be done with them.

MATLAB AND OCTAVE HAVE THE added advantage of incorporating a number of ready-made solutions for the visualisation of data. Chapter 4 is dedicated to explaining the plotting capabilities of the software as well as the formatting of the plots generated. In Chapter 5 we bring all the elements discussed and explain how MATLAB and Octave are powerful programming environments. We present the ways in which the software deals with the flow of a programme and demonstrates the use of functions and procedures. Finally, the last chapter provides an opportunity to see the different topics discussed earlier in the book in use within the context of an application. We have chosen specific examples from different areas in mathematics, engineering, finance and physics and their aim is to give the reader an idea of some of the things that can be achieved with MATLAB and Octave rather than provide a rigorous discussion about each subject.

THE BOOK WAS MADE POSSIBLE thanks to discussions with students and colleagues who offered suggestions for improvement. I am very grateful for their help, in particular to Kuldeep Singh and Alan McCall for their comments and suggestions. I want to take this opportunity to thank my editor at CRC Press, Francesca McGowan, whose input has been invaluable; similarly many thanks go to the technical reviewers whose thorough comments were more than welcome. I would also like to express my gratitude to my family, in particular to Antony and Bowman for their patience and understanding throughout the writing of this book.

London, U.K.
Dr Jesús Rogel-Salazar
May 2014

About the Author

DR JESÚS ROGEL-SALAZAR IS a member of the School of Physics, Astronomy and Mathematics at the University of Hertfordshire, UK, and a visiting researcher at the Department of Physics at Imperial College London, UK. He obtained his doctorate in Physics at Imperial College London for work on quantum atom optics and ultra-cold matter. He has held a position as senior lecturer in mathematics as well as a consultant in the financial industry since 2006. His interests include mathematical modelling, data science and optimisation in a wide range of applications including optics, quantum mechanics, data journalism and finance.

1

MATLAB® and Octave: The Essential Essentials

THE USE OF COMPUTERS HAS, without a doubt, changed the way in which science, engineering and many other disciplines compile and analyse data. Computers provide us with the ability of processing data at speeds that would be impossible with pen and paper alone, and it is thus of paramount importance to be able to instruct hardware and software to carry out the operations we require them to compute. This book deals with the way one can achieve these tasks with the help of MATLAB and Octave, two very similar numerical computing environments widely used by scientists and engineers.

1.1 MATLAB and Octave

MATLAB STANDS FOR *MATRIX LABORATORY* and is distributed by The MathWorks[1], originally created to facilitate matrix operations. It now includes capabilities to plot functions and data, creation of graphical user interfaces (GUIs) and even interaction with other programming languages such as C++ or FORTRAN.

[1] The MathWorks - http://www.mathworks.com

Octave, or to use its full name, GNU Octave[2], was developed as a convenient command line tool for solving problems numerically in a language that is mostly compatible with MATLAB. MATLAB is a trademark of The MathWorks and as such is a proprietary software subject to licensing. Conversely, GNU Octave shares many of the capabilities of MATLAB with the added merit of being freely distributed under the GNU General Public License. In this book we present code that is compatible with both languages in the expectation of enhancing the numerical computing capabilities of the reader.

[2] GNU Octave - http://www.octave.org

This book presents some of the most essential aspects of MATLAB and Octave programming, and it is important to mention that it is not meant to be an exhaustive manual for either. In cases where further in-depth manual-style information is required, the interested reader is recommended to consult directly MATLAB manuals[3,4] as well as Octave ones[5,6].

1.1.1   Obtaining MATLAB

MATLAB LICENSES CAN BE PURCHASED directly from The MathWorks, which shall be able to provide further information regarding the platforms supported as well as the system requirements for the software. The software is typically available for Windows, Mac OS and UNIX/Linux. Generally speaking, the installation procedure is guided by the software setup and a MathWorks account as well as an activation key is required. Further information can be obtained from www.mathworks.com.

[3] Higham, D. J. and N. J. Higham (2005). *MATLAB Guide*. Society for Industrial and Applied Mathematics
[4] Palm, W. J. (2008). *A Concise Introduction to MATLAB*. McGraw-Hill Higher Education
[5] Hansen, J. S. (2011). *GNU Octave Beginner's Guide*. Learn by doing: less theory, more results. Packt Publishing, Limited
[6] Eaton, J. W., D. Bateman, and S. Hauberg (2008). *GNU Octave Manual: Version 3*. A GNU manual. Network Theory Limited

1.1.2   Obtaining Octave

AS MENTIONED ABOVE, OCTAVE IS distributed under the terms of the GNU General Public License[7] and although it is primarily designed to run under Linux, there are implementations that run in Windows and Mac OS. The software can be directly downloaded from http://www.octave.org but it may be easier for the unexperienced user to download a packaged installation file. These packages are available for Windows and Mac OS and can be found in the Octave-Forge website: http://octave.sourceforge.net. The installation procedure is then guided by the package itself.

[7] GNU (June 29, 2007). General Public License, Free Software Foundation, version 3. http://www.gnu.org/licenses/gpl (Last visited Aug 4,2014)

1.2   Starting Up and Closing Down

1.2.1   Windows Systems

ON WINDOWS SYSTEMS MATLAB IS started by double-clicking the MATLAB icon on the desktop. This will open up a window divided in various subwindows and at the top the user will find the typical toolbars used in many Windows applications. One of the subwindows shows a command line terminal where the prompt is indicated with >>. It is in this window where we can type various commands to carry out calculations. To close MATLAB simply type

```
> quit
```

in the command line or close the window as you would do with any other application in Windows.

Octave can be launched in a very similar fashion; the main difference is that, unlike MATLAB, the only window that is available to the user is the command line, typically indicated with `octave:1>`. For the purposes of this book we will denote the command line for both MATLAB and Octave as `>` only.

To close Octave you can type `quit` as it is done for MATLAB or simply

```
> exit
```
Octave

It may seem very daunting at first not to have the comfort of icons around the main window provided by the Java GUI that comes with MATLAB, but do not let this put you off. In this chapter, as well as in the following ones, you will familiarise yourself with the command line environment and in no time you will master the use of MATLAB and Octave even if the familiar-looking icons are not there.

1.2.2   UNIX

IF YOU ARE A UNIX user you are probably familiar with the idea that a number of commands are entered directly in a terminal shell. To start MATLAB on a UNIX platform simply type the command `matlab` in the shell as follows:

```
> matlab
```
MATLAB

It may be the case that you will need to launch the software in an xterm window, depending on the version of MATLAB you are using. For further information about this please consult The MathWorks[8].

[8] The MathWorks - http://www.mathworks.com

Octave can be started in the same way; in other words, simply type the command `octave` at the system prompt as follows:

```
> octave
```
Octave

You will notice that, once again, MATLAB opens up a desktop with icons, whereas Octave stays as a terminal shell. Should you wish to start up MATLAB without the desktop you can issue the following command:

```
> matlab -nodesktop
```
MATLAB

and both Octave and MATLAB will look very similar. You can quit both MATLAB and Octave with the command quit.

1.2.3 Mac OS Systems

To START BOTH MATLAB AND Octave in a Mac OS environment you can double click the icons that appear in the /Applications folder after installation. Once again, MATLAB will start up with a desktop, whereas Octave will simply open in a terminal shell. Depending on your version of MATLAB you may have to start an X11 or XQuartz terminal depending on the version of MATLAB you have and your operating system. For more information about this please refer directly to the information provided by The MathWorks[9].

[9] The MathWorks - http://www.mathworks.com

It is possible to start the software directly from a terminal in a similar way to that of the UNIX system, i.e. by using the following commands: for MATLAB simply type the following command in the shell terminal:

```
> matlab
```
MATLAB

Please note that it may be necessary to modify the path in your computer so that the command line knows where to find the appropriate application.

In the case of Octave, the application can be started by typing the following in the command line:

```
> octave
```
Octave

Finally, to exit both MATLAB and Octave simply type `quit` at the command line:

```
> quit
```

1.2.4   Command Line Help

THERE ARE A NUMBER OF times when further information about a command or function is needed. MATLAB and Octave are able to provide some help from the command line prompt. Type `help help` (yes, twice!) for a brief synopsis of the help system. In MATLAB, typing `help` returns a list of topics:

```
> help

HELP topics:

matlab/general-General purpose commands.
matlab/ops  - Operators and special characters.
matlab/elmat - Elementary matrices and matrix...
```
MATLAB

In this case we have truncated lines and denoted them with ellipsis (. . .) as shown above.

In Octave, help explains the use of this function. We can obtain help for a particular command or function by preceding its name with the command help:

```
> help

For help with individual commands and functions
type
   help NAME
 (replace NAME with the name of the command
 or function you would  like to learn more
 about).
```

Octave

Let us take a look at the help for one of the built-in functions in the software, namely, the function ones:

```
> help ones

 ONES   Ones array.

 ONES(N) is an N-by-N matrix of ones.

 ONES(M,N) or ONES([M,N]) is an M-by-N
    matrix of ones
```

MATLAB

You can do the same in Octave. For certain functions, the information given by help can be quite lengthy. In order to see the information one screen at a time, first turn on the command more, i.e.,

```
> more on
> help ones
```

You can then hit any key to read on the information provided.

1.2.5   Demos in MATLAB

DEMONSTRATIONS CAN BE VERY USEFUL as they provide
some examples of the capabilities of the software. MATLAB
has a number of them and they can be accessed by typing
the command demo in the command line of the programme.
Please note that this command will clear any information
that is currently in MATLAB's workspace. Unfortunately
there is not an equivalent of this command in Octave, but
you can obtain a number of examples on the web or keep on
reading this book!

1.3   Using MATLAB and Octave as a Calculator

THE BASIC ARITHMETIC OPERATORS ARE addition +, substraction −, multiplication ∗, division / and exponentiation
^ and these are used in conjunction with brackets: ( ).

The symbol ^ is used to get exponents (powers): for example, 4^2 can be obtained with the commands 4^2=16. The
calculations can be typed directly in the command line:

```
> 9+2/8*6

ans =
    10.5000
```

There could be a certain amount of ambiguity with the
calculation above. Is this 9+2/(8∗6) or 9+(2/8)∗6? MATLAB
works according to the following priorities:

1. Quantities in brackets

2. Powers (1+4^2=1+16=17)

Priorities for arithmetic calculations.

3. The operators * /, working left to right (2/8*6=0.25*6)

4. The operators + −, working left to right (4+6−6=10−6)

The calculation we referred to earlier on is thus for 9 + (2/8)*6, by priority 3.

1.4 Numbers and Formats

MATLAB AND OCTAVE RECOGNISE SEVERAL different kinds of numbers, some of which are shown in Table 1.1. Similarly, the software can display numbers in different formats. They can be controlled with the command format and some common formats are shown in Table 1.2.

Type	Examples
Integer	2, −98765
Real	5.4321, −80.768
Complex	$7.76 − 6.42i$ ($i = \sqrt{-1}$)
Inf	Infinity (result of dividing by zero)
NaN	Not a Number, 0/0

Table 1.1: Some of the types that are supported in MATLAB and Octave.

The "e" notation is used for very large or very small numbers:

$$7.93796e+03 \ = \ 7.93796 \times 10^3 = \ 7937.96$$
$$7.93796e − 01 = 7.93796 \times 10^{-1} = 0.793796$$

ALL COMPUTATIONS IN MATLAB AND Octave are done in double precision, that is to say that about 15 significant figures are used. The displayed format of the output is handled by the command format; if the format is changed and you want to go back to the default simply use the command with no further input argument. A common formatting is the one given by

```
format compact
```

which suppresses blank lines in the output and allows you a
better use of the workspace real state.

Command	Example of output
format short	6.022 (4 decimal places)
format short e	6.022e+23
format long e	6.0221415e+23
format bank	6.02 (2 decimal places)

Table 1.2: Some number formats used by MATLAB and Octave.

1.5   Variables

So far both MATLAB and Octave have been used as a
simple calculator where the input is followed immediately
by the output. If we need to use the result of a calculation
in a later procedure, we can simply type the value returned.
However, as in a calculator, the software is capable of storing these values and provide us with a more efficient way
of carrying out calculations. One way to do this is to use
variables. Take a look at the code below:

```
> 1+4^2

ans =
    17


> ans*4

ans =
    68
```

The variable ans stores the value of the latest calculation performed.

The result of the first calculation is labelled ans by the
software and is used in the second calculation where its

value is changed. Please note that ans will always store the result of the latest calculation and therefore it is not recommended to use it at length. It is preferable to create ad hoc names to store values. In that way we can assign the results of calculations to variables that can later be retrieved. For example, we can assign the result of $1 + 4^2$ to the variable x as follows:

```
> x=1+4^2

x=
    17
```

Variable assignation is done with the = sign.

In this case x stores the value 17. Let us look at another example:

```
> y = x*4

y=
    68
```

Here $y = 4x = 4(17) = 68$. The value held by x will only be changed whenever we explicitly operate on it and therefore it can be safer to use in later calculations. If we were to change the value of x it would be necessary to recalculate y. These are examples of *assignment statements*: values are assigned to variables. Each variable must be assigned a value before it may be used on the right of an assignment statement.

The assignment of values to variables is done with the = operator; notice that the value to be assigned appears on the right of the operator.

1.5.1   *Variable Names*

LEGAL NAMES OF VARIABLES CONSIST of any combination of letters and digits, starting with a letter. The following names are allowed to be used as variables:

```
RateVal, Phi2y, x1, X2, z1y2, Theta_1
```

whereas the following names are **not** allowed to be used as variable names:

```
Rete-Val, 2y, %x, @sign, Alpha+1
```

The latter are not allowed because they conflict with normal syntax of commands used by the software. Notice that the use of underscores (_) is allowed in the declaration of variables, but not the dash. In general, it is recommended to use names that reflect the values that the variables represent.

Another thing to remember is that there are some special values that are defined in the software and thus these names should not be used as variables, as this can create conflicts difficult to debug. One example is that of the value of π, represented by pi=3.14159...; another one is the value for floating point relative accuracy EPS: eps = 2.2204e−16 for double precision. This number can be thought of as the smallest distance between two floating point numbers as represented by the precision of the computer used. For example, a computer with a given floating point relative accuracy eps cannot find a floating point number between 1 and 1 + eps. Similarly, if you require to do arithmetic calculations with complex numbers, both i and j have the value $\sqrt{-1}$; however, these two variables can indeed be redefined and care must therefore be taken:

The mathematical constant π is represented in the software by the constant pi.

The floating point relative accuracy EPS is represented by the constant eps.

```
> i, j, i=42

ans =
    0 + 1.0000i
    0 + 1.0000i
    i=42
```

In order to avoid this sort of situation, it is recommended to replace the complex i and j with 1i. The latter is understood both by MATLAB and Octave and it improves robustness. For example, we can define the complex number $a = 1 + 2i$ as

```
> a=1+2*1i

a =
    1.0000 + 2.0000i
```

To improve robustness, replace complex i and j with 1i.

We can also define a purely imaginary number, in other words one that has no real part, such as $b = 6i$:

```
> b=6*1i

b =
    0.0000 + 6.0000i
```

The usual arithmetic operations that we have discussed can also be applied to complex numbers in the software, for example:

```
> a+b

ans =
    1.0000 + 8.0000i
```

1.6   Suppressing Output

IN MANY CASES IT IS useful to see the result of every calculation we perform in MATLAB and Octave; however, there may be many situations where either we do not need to see the result, or the output is so large that it is not practical to display it. In those cases, we can suppress the output of commands by appending a semicolon (;) at the end of the command. For instance, issuing the command

Use a semicolon (;) to avoid displaying output.

```
> x=-5

x =

    -5
```

tells the software to display the result of the operation, in this case echoing the assignment of -5 to the variable x.

Let us contrast this behaviour with the use of the semicolon at the end of the line to suppress the output. Imagine that we have a parameter x for which the value is known and thus we do not need to display it. However, we are interested to know the output of calculations made with that parameter. For example:

```
> x=-5; y = 9*x, z = x^3+y

y =

    -45
z =

    -170
```

In the code above we have suppressed the displaying of x, however since the other commands are separated by a comma (,) and nothing at the end of the line, the values of y and z are displayed.

1.7   Built-In Functions

1.7.1   Trigonometric Functions

TRIGONOMETRIC FUNCTIONS ARE SO WIDELY used it is not surprising that MATLAB and Octave have them built-in. In order to call the most common trigonometric functions we need to use sin, cos and tan, and their arguments should be in radians.

```
> x = 3*cos(pi/4), y = 16*sin(pi/3)


x =

   2.1213
y =
   13.8564
```

The common trigonometric functions are sine (sin), cosine (cos) and tangent (tan).

Similarly, it is useful to have the inverse trigonometric functions. These can be invoked with asin, acos and atan and their output is in radians. For example:

The inverse trigonometric functions are asin, acos and atan.

```
> acos(x/3), asin(y/16)


ans =
   0.78540
ans =
   1.0472
```

The values shown above correspond to numerical values of $\pi/4$ and $\pi/3$.

1.7.2   Other Elementary Functions

HAVING DEFINITIONS FOR THE TRIGONOMETRIC functions is very useful and indeed necessary, but we need access to a wider range of functions and procedures. Some may include functions such as the square root, the exponential function and the logarithm.

These functions can be called with sqrt, exp, log and log10. It is important to mention that the inverse of the exponential function $\exp(x) = e^x$ is denoted by log in MATLAB and Octave.

Other common functions are sqrt, exp, log and log10.

Function	MATLAB and Octave
$\sin(x)$	sin(x)
$\cos(x)$	cos(x)
$\tan(x)$	tan(x)
$\sqrt{x}$	sqrt(x)
e^x	exp(x)
$\ln(x)$	log(x)
$\log_{10}(x)$	log10(x)

Table 1.3: Some of the mathematical functions defined in MATLAB and Octave.

Many other functions are defined in both MATLAB and Octave as discussed later on in Section 5.6, but let us see some examples here.

Given a real number $x = 8$ for instance, we can calculate:

- $\sqrt{8+1} = 3$

- $\exp(8) = 2.981 \times 10^3$

- $\cos(\ln(8)) = -0.487$

- $\log_{10}(8^3) = 2.7093$

which can be carried out with the following commands in MATLAB and Octave:

```
> x=8;
> sqrt(x+1), exp(x), cos(log(x)), log10(x^3)

ans =

    3
ans =

    2.9810e+03
ans =

   -0.4870
ans =

    2.7093
```

1.8   Characters, String and Text

USING THE SOFTWARE AS A calculator is fine, but in order
to make it more flexible there is the need to process text as
well as numbers. This gives us flexibility for interacting with
the software in terms of input and output both in the screen
and computer files. In order to manage text, MATLAB
and Octave use a "character" datatype, which allows us to
define some text as a string stored in a vector or array of
characters. We shall discuss vectors and arrays in Chapters 2
and 3.

For example, we can indicate to the software that we need
to store the character *t* in a variable instead of a number as
we have done until now. This can be done as follows:

```
> str1 = 't'
```

Note that the string is enclosed
between apostrophes or single
quotes: 't'.

This command will assign the character *t* to the 1-by-1
character array str1. It is important to mention that the
use of apostrophes or single quotes is common to both

MATLAB and Octave and thus we encourage this use to improve the portability of code between both languages. Nonetheless, it may be useful to know that Octave also supports the use of quotation marks to define strings. In this case the following command is valid in Octave:

Octave supports the use of quotation marks to define strings.

```
> str = ''t''
```

Octave

Let us have a look at further examples: the assignment,

```
> str2 = 'abc'
```

assigns the characters *abc* to the 1-by-3 character array str2. Strings can be combined by using the operations for array manipulations that will be explained in Chapters 2 and 3.

For instance, we can concatenate the two strings defined above as follows:

```
> str3 =[str1,str2]

ans =
    tabc
```

The result of concatenating two strings is a new string.

In the command shown above we have assigned the characters "tabc" to the 1-by-4 character array str3. Notice that there is no space between the elements of the new string. We can add a space as follows:

```
> str3=[str1, ' ', str2]

ans =
    t abc
```

where we have effectively concatenated three strings: str1, str2 and a string given by a space.

The assignment shown in the following using MATLAB example:

```
> str4 = [str3, ' is a string'; ...
         '        of letters']


 str4 =

    tabc is a string
           of letters
```

MATLAB

Concatenation in MATLAB only works if the strings have the same number of elements. That is why we have left the blank space in the second string used in this example.

assigns the value "tabc is a string" and "   of letters" to the 2-by-16 character array str4. Please note that the second string starts with a number of leading blank spaces. This is because the number of characters of both rows in the array needs to have the same number of elements. In the code above, the ellipsis (. . .) indicates to the software that the command is continued on the following line.

The ellipsis, . . ., is used to split commands into several lines.

However, in Octave, the software is able to pad the strings to force them into arrays of equal length so the code above can be written as:

```
> str4 = [str3, ' is a string';
         'of letters']


 str4 =
    tabc is a string
    of letters
```

Octave

Concatenation in Octave works for strings of different lengths as the software pads the strings to have the same number of elements.

where Octave has added blank spaces at the end of the second line to get a consistent array. Finally, note that the ellipsis is not needed in Octave.

1.8.1   *Comparing Strings*

WE ARE FAMILIAR WITH THE idea of comparing numbers to decide if one is larger, smaller or equal to another. In the case of strings, we can think of checking if two strings are equal, for example, if a script requires the user to type an answer such as 'Continue' or 'Stop'.

This can be achieved with the help of the strcmp command that takes as input two strings (or variables of string type) separated by a comma. Two strings are considered to be equal if they have the same length and the content is the same, including the case in which the strings are written. Let us take a look at an example: If we compare the strings 'Continue' and 'Stop' the result is false as they do not have the same length or content. This is denoted as a zero (0) in MATLAB and Octave:

We can compare two strings with the help of the strcmp command.

```
> s1='Continue'
> s2='Stop'
> strcmp(s1,s2)

ans =
      0
```

Nonetheless, if we compare a string s1 that contains the word *Continue* with the string 'Continue' the result is true, denoted as a one (1):

```
> strcmp(s1,'Continue')

ans =
      1
```

1.8.2   Converting Strings to Values

SOMETIMES IT IS CONVENIENT OR even necessary to
convert a character to the corresponding number, or vice
versa. These conversions are accomplished by the commands str2num - which converts a string to the corresponding number, and two functions, int2str and num2str,
which convert, respectively, an integer and a real number to the corresponding character string. These commands are useful for producing titles and strings, such
as: The value of pi/4 is 0.7854. This can be generated by
the following command:

> To convert between strings and
> values use commands such as
> num2str and int2str.

```
> ['The value of pi/4 is ', num2str(pi/4)]
```

The same syntax can be used to display the value of variables. Take a look at the following example:

```
> Universe = 42; u = 1/42;

> ['The value of Universe is ', ...
    int2str(Universe),...
    ', u = ' , num2str(u)]

ans =
    The value of Universe is 42, u = 0.02381
```

1.9   Saving a Session

AFTER USING MATLAB AND OCTAVE to perform a number
of calculations or procedures, you may want to save some of
the results and store them for later retrieval. The software

allows us to save the variables that are in the memory into a file; this can be done as follows:

```
> save filename.mat
```

Use save to save the variables in your workspace to a file.

This will save the current values of all variables to a file called `filename.mat`; it is a good practice to use a name for the file that provides some useful information regarding its contents. It is important to note that the file that is generated in this way can only be read and edited by either MATLAB or Octave. If you need to store the information in a formatted text file or in a binary file, please refer to Section 5.10.

In MATLAB it is possible to leave the extension .mat out from the save command. It is however expected by Octave.

Once we have saved the file as mentioned above, we can retrieve the information as follows:

```
> load filename
```

Use the command load to retrieve a saved session.

This command instructs MATLAB and Octave to bring up the variables and values that are stored in the file called `filename.mat`; once this is done we can use and modify their values.

SOMETIMES IT IS VERY USEFUL to check the existing variables in a workspace and list their names and sizes as well as the amount of memory they take and the class they belong to. This can easily be done with the whos command:

```
> whos
```

The command whos displays the existing variables in the current workspace.

This command will list a summary of the variable names, their size, the number of element as well as the bytes they occupy and their type. For instance:

```
> a=1; x=[1,3]; y=magic(5);

> whos

   Name        Size              Bytes  Class

   a           1x1                   8  double
   x           1x2                  16  double
   y           5x5                 200  double
```

In this case we have defined a variable a that holds a scalar, a variable x that holds a row vector with two elements, and a variable y that holds a 5×5 magic square. We can see from the summary above the memory (in bytes) used by each variable and their type, which in this case is double for all of them.

An $n \times n$ magic square is an arrangement of the integers $1, \ldots, n^2$ such that row sums, column sums and diagonal sums are all equal to the same value.

FINALLY, IT MAY BE OF importance to you to keep a record not only of the variables in memory, but also of the various commands and text that are used in a particular session. In this case we can create a journal or diary with the following command:

```
> diary filename
```

The command diary keeps a log of the input and output in a file.

This command instructs the software to keep track of all subsequent text that is sent to the screen and save it in a file called filename. In order to stop the recording we need to use the following command:

```
> diary off
```

1.10 Summary

IN THIS CHAPTER WE HAVE presented an overview of what
MATLAB and Octave are, as well as general guidelines on
how to obtain the software and how to start a session in
each of them.

We have covered some useful commands (such as `help`
and `save` and `whos`) as well as some general features of the
software environment. We are now able to understand the
format in which numbers are presented as well as define
variables to store the output of calculations. Similarly, we
have introduced some common built-in functions such as
`sin`, `cos`, `tan`, `log` and `exp`.

We have seen how MATLAB and Octave are capable of
dealing with various types of numbers including integers,
decimals and complex numbers. In this way, MATLAB and
Octave can be used in the same way as a calculator. In the
next chapter we will see how the software can be used to
define vectors and how to operate with them.

1.11   Exercises

1. MATLAB and Octave have a built-in function called clc. Request the software for help and information about this function.

2. Using the command-line in MATLAB and Octave, calculate the following expressions:

   (a) $\frac{8.2^3 - 6^2}{\sqrt{2}} + 10\,(3\pi)$

   (b) $9\left(\sqrt[3]{4}\right)\left(\frac{0.856}{3}\right)$

   (c) $5.3\left(42(1.2^8)\right)\left(\frac{1}{\sqrt{(1-0.8^2)(1+0.8^2)}}\right)$

3. Find out what the following functions do:

   (a) sind

   (b) cosd

   (c) tand

4. Using MATLAB and Octave, calculate the value of the following expressions:

   (a) $\frac{\sin(3.5\pi)}{\cos(1.8\pi)} + \frac{1}{2}\cot(2.1\pi)$

   (b) $6\sin\left(75°\right) + \frac{1}{4}\cos\left(63°\right)$

   (c) $\log_{10}(4987) + \ln(6.5) - \tan(4.1\pi)$

5. Consider the following strings and identify which ones are valid names for variables in MATLAB and Octave:

   (a) First_One

   (b) velocity 1

   (c) sin

   (d) myv@riable

   (e) MyCalculation

   (f) Last-place

(g) whichOne

(h) i

6. If $t = 10$, $a = 9.5 - 5i$, $f = \exp(0.1)$ and $b = -7.356$, what are the values of the following expressions? Use MATLAB and Octave to calculate the answers:

(a) $q = t^2 + b$

(b) $w = f - \frac{b}{t}$

(c) $e = a^3 - f$

(d) $r = f^{-t} + \frac{6}{a}$

7. For the evaluations carried out in Exercise 6, create strings that display their value using the following message:

The value of *expression* is *value*,

where *expression* is each of the expressions in Exercise 6 and *value* is the calculated result.

8. Use the number format commands in MATLAB and Octave to display the following calculations using the default number of decimal places, 15 decimal places and in scientific notation:

(a) $\frac{4}{7} + 5.1^3$

(b) $\exp(0.3) - \frac{1}{8.1}$

(c) $\sin\left(\frac{3\pi}{7}\right)$

9. Write a series of commands that enable you to find the roots of a quadratic equation $ax^2 + bx + c = 0$. Set up the commands to solve for $a = 3$, $b = 5$ and $c = -6$.

10. What does the built-in function clear do?

2

Vectors and Vector Operators

IN CHAPTER 1 WE COVERED some of the most basic aspects of MATLAB® and Octave, and have seen how the software can be used as a simple calculator to do arithmetics, as well as performing operations using built-in functions. Nonetheless, both MATLAB and Octave are far more powerful than that, and to unlock this power it is important to familiarise ourselves with the objects that the software is able to manipulate. In this chapter we will see how to deal with vectors and the type of operations we can do with them.

2.1 Vectors

FOR THE PURPOSES OF USING MATLAB and Octave, we can think of a vector as an arrangement of elements in a column or a row. In that manner, vectors are effectively lists of numbers for example, separated by either commas or spaces. The number of entries is known as the "length" of the vector and the entries are often referred to as "elements" or "components" of the vector. The entries must be enclosed in square brackets. For example, to enter the vector

$$\mathbf{vector_1} = \left(3, 42, \sqrt{25}\right) \qquad (2.1)$$

into MATLAB and Octave we type the following in the command line:

```
> vector1 = [3 42, sqrt(25)]

vector1 =

   3    42    5
```

Use square brackets to define a vector.

In the example above we have left a space between the first element (with value 3) and second one (with value 42). We could have done this explicitly with a comma, as shown in the separation of the second (42) and third ($\sqrt{25}$) elements.

The number of elements in a vector can be separated with a blank space or with a comma (,).

We must be careful with the use of spaces as they can change the number of elements in a vector and their value. Let us have a look at the following example:

```
> vector2 = [8+9 5-4]

vector2 =
   17    1


> vector3 = [8+9 5 -4]

vector3 =
   17    5    -4
```

Be careful when using blank spaces to define vectors; they can change the number of elements and their value.

In the first line we have defined a row vector, vector2, with two elements, but the extra space in the assignment of vector3 creates a row vector with three elements instead.

IF WE DEFINE TWO VECTORS of the same length, we can carry out arithmetic operations with them such as addition and substraction element by element. Let us consider the vectors vector1 and vector3 defined above as they are both of length 3. For example, given the vector $\mathbf{vector_1}$ shown in Equation (2.1) and the vector

$$\mathbf{vector_3} = (17, 5, -4),\qquad\qquad(2.2)$$

we can add them together as follows:

```
> vector1 + vector3

ans =
   20    47    1
```

Note that we can use the vectors defined above to construct new vectors and carry out addition/substraction with them. Take a look at the following examples. We can create a new vector:

$$\mathbf{vector_4} = 50 \times \mathbf{vector_1} = (150, 2100, 250).\qquad(2.3)$$

This can be done in MATLAB and Octave as follows:

```
> vector4 = 50*vector1

vector4 =
   150    2100    250
```

Similarly, we can construct other vectors using the same pattern:

```
> vector5 = 3*vector1 - 2*vector3

vector5 =
   -25   116    23
```

As we can see in the first example above, a vector can be multiplied by a scalar (i.e. a number), or added/subtracted to another vector of the same length. The operations are carried out element by element.

Arithmetic operations such as addition or substraction can be done with vectors of the same length.

Let us see what happens when the length is not the same:

```
> vector1 + vector2

   ??? Error using ==> plus Matrix dimensions
   must agree.
```

MATLAB

```
> vector1 + vector2

   error: operator +: nonconformant
   arguments (op1 is 1x3, op2 is 1x2)
```

Octave

An error is returned by both MATLAB and Octave because vector1 and vector2 have different lengths. This can easily be seen from the error messages returned by the software. In the case of MATLAB we are told that the dimensions of the vectors must agree, whereas Octave explicitly tell us that one vector is 1×3 and the other one is 1×2. From this point of view vectors can be seen as arrays and we shall discuss this idea in Chapter 3.

Once we have defined a few vectors it is possible to construct more vectors from the existing ones. Consider the following examples: let us define the vectors

$$v_1 = (9, 8, 7), \qquad (2.4)$$
$$v_2 = (6, 5). \qquad (2.5)$$

With these two vectors, we can envisage to build a new one, v_3, such that the first elements are three times those of vector v_1 and the last elements are twice those of vector v_2. This can easily be done in MATLAB and Octave as follows:

```
> v1=[9 8 7]; v2=[6 5];
> v3=[3*v1,-2*v2]

v3 =
    27    24    21   -12   -10

> sort(v3)

ans =
   -12   -10    21    24    27
```

The sort command can be used to sort the elements of an array.

After defining the vectors v1 and v2 we have concatenated them to build a new vector; this is similar to what we did with strings in Section 1.8. In the last line we have used one of the built-in functions to sort the elements of the new vector v3 in ascending order. Other built-in functions are presented in Section 5.6.

Vectors can be concatenated using the same syntax for defining vectors; the difference is that now the elements are vectors themselves.

SINCE EACH VECTOR HAS A determined number of elements, it is convenient to be able to operate not only on the vector itself, but also on selected elements individually. This can be done by using an index that refers to the position of the element entry. For example, given the vector

$$v_3 = (27, 24, 21, -12, -10), \qquad (2.6)$$

calculated above, we can retrieve the first element, i.e. 27, as follows:

```
> v3

v3 =
    27    24    21    -12    -10


> v3(1)


ans =
    27
```

Each element in a vector can be addressed by an index starting from number 1.

The elements start being enumerated from 1, so in the example above v3(1) refers to element number 1 of the vector v3.

We can also reassign values; for instance, if we require the third of element v3 to be zero we can do the following:

```
> v3(3)=0

v3 =
    27    24    0    -12    -10
```

Here we are reassigning the value of the third element to zero, while leaving the other elements intact.

2.2   The Colon Notation (:)

IN THE PREVIOUS SECTION WE have seen how to define a vector as a list of numbers. We can enter the list directly in the command line, but as the number of elements increases, this method becomes very unpractical. Another

way of defining a vector is by using a colon (:) to specify the beginning and ending elements of the vector as
beginning:ending. This will instruct the software to create
a vector that starts with beginning, and adds 1 successively
until ending (but not beyond) is reached:

The colon notation allows us to define vectors by specifying a start and an end as beginning:ending.

```
> [3:9.6]

ans =
   3   4   5   6   7   8   9

> [2:7]

ans =
   2   3   4   5   6   7
```

In the first line we have asked the software to list numbers
starting with 3 and ending with 9.6. Remember that the sequence cannot go beyond ending. In this case the sequence
is truncated at 9. In the second example above we are listing
numbers starting with 2 and finishing with 7. In this case
the sequence reaches the ending point and no truncation is
needed.

If we try to create a series where the ending cannot be
reached by adding 1 successively an empty vector will be
returned. This is the case, for instance, when ending is
smaller than beginning. In the case of MATLAB we obtain
the following error message:

MATLAB

```
> [2:1]

ans =
   Empty matrix: 1-by-0
```

whereas in Octave the following is displayed:

```
> [2:1]

ans =
     [](1x0)
```

A MORE GENERAL WAY OF creating a vector is by defining a series to specify not only the starting and ending values, but also a step: beginning:step:ending. This produces a vector of entries which starts with the value of beginning, incrementing by the value of step until reaching the value of ending (but not going beyond it). Note that the step can be a negative number.

beginning:step:ending can be used to define a vector.

```
> 0.5:0.1:1.0

ans =
     0.5    0.6    0.7    0.8    0.9    1.0

> [-10:-1:-15]

ans =
     -10   -11   -12   -13   -14   -15
```

With a negative step it is possible to have decreasing sequences.

2.3   Extracting Parts of a Vector

NOW THAT WE KNOW HOW to construct a vector, either by listing its entries or by specifying a sequence, we can turn our attention to the way of extracting a part of that vector. One way to do this is by using the index of each element as explained at the beginning of this chapter. However, when a large number of elements are involved that method may not be practical.

Nonetheless, by combining the use of the colon notation as shown in Section 2.2 with the indexing of elements, we can extract portions of a vector in a more effective manner. Let us look at an example:

Combining the colon notation (:) with indexing is a powerful way to extract parts of a vector.

```
> v4 = [0:1:3, -10:2:-6]

v4 =
    0   1   2   3  -10  -8  -6
```

In the example above we have defined a vector v4 from two sequences using the colon notation. We can then obtain elements 3 to 6 using once again the colon notation as follows:

```
> v4(3:6)

ans =
    2   3  -10  -8
```

Note that we are using round brackets to refer to indices.

If we are interested in the first, third and fifth elements we can obtain them as follows:

```
> v4(1:2:6)

ans =
    0   2  -10
```

We would have obtained the same answer if we had used the following command:

```
> v4(1:2:5)

ans =

   0    2   -10
```

This is because the series cannot go beyond the ending value of the series specified by the colon notation as explained above.

It is important to mention that the arguments used to access the elements of a vector are in fact arrays themselves. This can be easily seen with the following example. Let us define a couple of vectors as follows:

The arguments used to access the elements of an array are arrays themselves.

```
> a=[1:3, -8:2:2, 2:-2]

a =

   1   2   3  -8  -6  -4  -2   0   2


> b=[1 5 8:9]

b =

   1   5   8   9
```

We can now subset vector a using the elements of vector b as follows:

```
> a(b)

ans =

   1  -6   0   2
```

In this case we are accessing the first, fifth, eighth and ninth elements of vector a.

2.4   Column Vectors

THE VECTORS USED IN THE previous section are called *row vectors* because the software accommodates the elements one by one in a row. In this section we will see how to instruct the software to define a column vector, i.e. a vector whose entries are aligned in a column. A column vector requires a new line to be inserted after each element; we achieve this simply by separating the elements with a semicolon (;). Consider the vector

Row vectors organise the elements in a row. Column vectors are organised in separate lines, i.e. columns.

$$\textbf{cvector}_1 = \begin{pmatrix} 3 \\ 42 \\ \sqrt{25} \end{pmatrix}. \tag{2.7}$$

We can enter this in the software by typing the following in the command line:

```
> cvector1 = [3; 42; sqrt(25)]

cvector1 =

   3

   42

   5
```

A semicolon (;) is used to define columns.

Let us have a look at other examples:

$$\textbf{cvector}_2 \;\; = \;\; \begin{pmatrix} 8+9 \\ 5 \\ -4 \end{pmatrix}, \tag{2.8}$$

$$\textbf{cvector}_3 \;\; = \;\; -\textbf{cvector}_1 + (4)\,\textbf{cvector}_2. \tag{2.9}$$

The vectors above can be created as follows: for the first one we have

```
> cvector2 = [8+9; 5; -4]

cvector2 =

   17

    5

   -4
```

whereas for the second one

```
> cvector3=-1*cvector1 + 4*cvector2

cvector3 =

   65

  -22

  -21
```

As we can see from the example above, column vectors can also be used to carry out addition and substraction, provided they all have the same length.

2.5   *Transposition of Vectors*

IT IS QUITE CLEAR TO see that a row vector can easily be transformed into a column vector and vice versa. In mathematical terms this operation is called a transposition. In MATLAB and Octave we can transpose a vector by appending an apostrophe or single quote (') to the name of the variable that holds the vector. For example, given the vector

A vector can be transposed with the use of an apostrophe (').

$$\mathbf{r} = (10, 9, 8),\qquad(2.10)$$

its transpose would be given by

$$\mathbf{r^T} = \begin{pmatrix} 10 \\ 9 \\ 8 \end{pmatrix}. \qquad (2.11)$$

This can be calculated in MATLAB and Octave as follows:
Let us enter the vector **r**:

```
> r=[10 9 8]

r =

    10    9    8
```

We can now obtain the transpose by typing the following:

```
> r'

ans =
    10
     9
     8
```

Transposition transforms a row-vector into a column-vector, and vice-versa.

The opposite transformation can also be carried out; in other words, given a column vector we can carry out a transformation to obtain a row vector.

For instance, given the column vector

$$\mathbf{s} = \begin{pmatrix} 1 \\ 3 \\ 5 \end{pmatrix}, \qquad (2.12)$$

its transpose would be

$$\mathbf{s^T} = (1,3,5) \qquad (2.13)$$

and in MATLAB and Octave this can be obtained as follows:

```
> s=[1; 3; 5], s'

s =
  1
  3
  5

ans =
  1   3   5
```

In the examples above we started out defining a row vector
r which is then easily converted into a column vector by
appending the apostrophe to the variable name, i.e. r'.
Similarly, the column vector s is transposed by typing s'.

As we have seen above, if we want to carry out addition and
substraction with vectors they need to have the same length.
They also need to be of the same kind, in other words they
all need to be either column or row vectors. Let us have a
look at an example; in MATLAB if we tried to add 3 times
the row vector r to 4 times the column vector s defined
above we would obtain the following output:

```
> new_vector1 = 3*r + 4*s

??? Error using ==> plus
Matrix dimensions must agree
```

MATLAB

However, Octave *broadcasts* vectors, matrices and arrays until
the objects have compatible size in order for the operation
to be valid and the software issues a warning in this regard.

Octave is able to *broadcast*, i.e.
repeat, a smaller vector to a larger
one to carry out computations.

So the operation above would result in the following output
from Octave:

```
> new_vector1 = 3*r + 4*s

warning: operator +: automatic broadcasting ...
operation applied

ans =

   34   31   28
   42   39   36
   50   47   44
```

Octave

In versions of Octave prior to 3.6.0 the behaviour was similar to that of MATLAB:

```
> new_vector1 = 3*r + 4*s

error: operator +:
nonconformant arguments (op1 is 1x3, op2 is...
3x1)
```

Octave

We get an error because, although the vectors do have the
same length, they are not of the same kind, i.e. one is a row
vector whereas the other one is a column vector. In order for
the above operation to return a correct answer we need to
transpose one of the vectors. We can, for instance, transpose
vector s and therefore the final result is stored in a row
vector,

Some arithmetic operations (+ or
−) require the vectors to be of the
same kind.

```
> new_vector1 = 3*r + 4*s'

new_vector1 =
    34    39    44
```

or we can instead transpose vector r and the result is stored in a column vector

```
> new_vector2 = 3*r' + 4*s

new_vector2 =
    34
    39
    44
```

2.6   Vector Multiplication

IN SECTION 1.3 WE SAW some simple operators that work on scalars, i.e. numbers. In this section we will describe operators that act on vectors as defined in the previous section.

We shall take a look at two ways to understand the product of two vectors. In both cases the vectors concerned must have the same length.

2.7   Scalar Product, *

THE FIRST PRODUCT IS THE standard scalar product. Suppose that **u** and **v** are two vectors of length n, with **u** being a row vector and **v** a column vector:

$$\mathbf{u} = (u_1, u_2, \ldots, u_n), \qquad (2.14)$$

$$\mathbf{v} = \begin{pmatrix} v_1 \\ v_2 \\ \vdots \\ v_n \end{pmatrix}. \qquad (2.15)$$

The scalar product is defined by multiplying the corresponding elements together and adding the results to give a single number, i.e. a scalar:

$$\mathbf{u} * \mathbf{v} = \sum_{i=1}^{n} u_i v_i, \qquad (2.16)$$

Definition of the scalar product.

where we have used the symbol $*$ to denote the operation (this indeed is the symbol used by MATLAB and Octave for the scalar product).

For example, let

$$\mathbf{u} = (10, 5, 0), \qquad (2.17)$$

$$\mathbf{v} = \begin{pmatrix} 2 \\ 4 \\ -6 \end{pmatrix}, \qquad (2.18)$$

then $n = 3$ and

$$\mathbf{u} * \mathbf{v} = (10 \times 2) + (5 \times 4) + (0 \times -6) = 40. \qquad (2.19)$$

This product can be carried out in MATLAB and Octave as follows:

```
> u = [ 10, 5, 0]; v = [2; 4; -6];
> prod = u*v        % row times column vector


ans =

    40
```

The scalar product of two vectors is denoted by * in the software. The text followed by the % symbol denotes a comment.

Note that the elements in the first vector above are separated by commas (row vector) and the ones for the second vector by semicolons (column vector). This is important as the multiplication can only be done with the appropriate dimensions. Let us consider a new vector

$$\mathbf{w} = (1,2,3), \tag{2.20}$$

which can be entered into the software as

```
> w=[1 2 3]

w =

   1   2   3
```

If we want to calculate $\mathbf{u} * \mathbf{w}$ we could try the following:

```
> u*w
   ??? Error using ==> * Inner matrix
   dimensions must agree.
```

MATLAB

```
> u*w
   error: operator *: nonconformant arguments
   (op1 is 1x3, op2 is 1x3)
```

Octave

An error results because $\mathbf{w}$ is not a column vector. Recall from Section 2.5 that transposing (with ') turns column vectors into row vectors and vice versa. So, to form the scalar product of two row vectors or two column vectors we can use the transposition to obtain a correct result:

```
> u*w'         % u & w are row vectors

ans =
    20
```

The scalar multiplication requires the vectors to have the same length and have the appropriate dimensions.

LET US NOW TAKE A look at a common application of the scalar product. We are familiar with the concept of the Euclidean length of an n-dimensional vector defined as the norm of a vector, denoted by $|\mathbf{u}|$:

$$|\mathbf{u}|^2 = \sum_{i=1}^{n} |u_i|^2. \qquad (2.21)$$

The Euclidean length can be calculated using the scalar product.

This can be computed with MATLAB and Octave in the following two ways:

```
> sqrt(u*u')

ans =
    11.180

> norm(u)

ans =
    11.180
```

The norm function calculates the Euclidean length of a vector.

In the first case we have taken the square root of the scalar product of the vector $\mathbf{u}$ with itself (and we have used the transposition operator). In the second case we have used a built-in function called `norm` that takes a vector as an input and returns its norm.

2.8   Dot-Star Product, .∗

ANOTHER WAY TO CONSTRUCT THE product of two vectors (of the same length) is what we will refer to as the dot-star product due to the symbols used in the software to carry out the operation. Unlike the scalar product introduced in the previous section, this one involves vectors of the same type, i.e. they all are row vectors or all column vectors. The result is a vector of the same length as the original ones and its components are element-by-element multiplications of the original two vectors. For instance, if **u** and **v** are two vectors of the same type, then the dot-star product is given by

> The dot-star product is an element-by-element multiplication of the two original vectors.

$$\mathbf{u} \circ \mathbf{v} = [u_1 v_1, u_2 v_2, \ldots, u_n v_n]. \qquad (2.22)$$

This operation is known as the Hadamard product and can be carried out in MATLAB and Octave with the .∗ operator. Using the vectors **u**, **v** and **w** defined in Section 2.7 we can calculate the following:

$$\begin{aligned} \mathbf{u} \circ \mathbf{w} &= (10, 5, 0) \circ (1, 2, 3), \\ &= (10, 10, 0). \qquad (2.23) \end{aligned}$$

```
> u.*w

ans =
   10   10   0
```

Similarly, the following operation can be carried out:

$$
\begin{aligned}
\mathbf{u} \circ \mathbf{v}^T &= (10, 5, 0) \circ (2, 4, -6), \\
&= (20, 20, 0), \qquad\qquad (2.24)
\end{aligned}
$$

which in MATLAB and Octave is calculated as follows:

```
> u.*v'

ans =
   20   20   0
```

A product of interest is the dot-star product of a vector with itself. This can easily be done as shown in the following two examples. For a row vector

$$
\begin{aligned}
\mathbf{u} \circ \mathbf{u} &= (10, 5, 0) \circ (10, 5, 0), \\
&= (100, 25, 0). \qquad\qquad (2.25)
\end{aligned}
$$

```
> u.*u

ans =
   100    25    0
```

The dot-star product of a row-vector with itself.

For a column vector,

$$
\begin{aligned}
\mathbf{v} \circ \mathbf{v} &= \begin{pmatrix} 2 \\ 4 \\ -6 \end{pmatrix} \circ \begin{pmatrix} 2 \\ 4 \\ -6 \end{pmatrix}, \\
&= \begin{pmatrix} 4 \\ 16 \\ 36 \end{pmatrix}. \qquad\qquad (2.26)
\end{aligned}
$$

```
> v.*v

ans =

     4

    16

    36
```

The dot-star product of a row-vector with itself.

In Section 2.10 we shall see another way to calculate this operation. In general, the dot-star product can be very useful in carrying out multiplications over lists of objects. It is important to note that the dot-star product over two vectors will result in another vector of the same length and dimensions of the original vectors.

The dot-star product results in a vector of the same length and dimensions of the original ones.

2.9   Dot-Division of Vectors, ./

MATHEMATICALLY, THE DIVISION OF ONE vector by another is not defined. Nonetheless, MATLAB and Octave have a shortcut operation to carry out an element-by-element division of two vectors, namely, ./, and can only be used for vectors of the same size and type.

The dot-division is an element-by-element division, similar in nature to the dot-star product.

```
> a=1:2:10, b=2:2:10

a =
    1    3    5    7    9
b =
    2    4    6    8    10

> a./b
ans =
    0.5000    0.7500    0.8333    0.8750    0.9000
```

Since division by zero is not defined, we must be careful when implementing this operation:

```
> a./(b-6)

ans =
  -0.2500  -1.5000    Inf   3.5000   2.2500

> c=-3:3

c =
  -3  -2  -1   0   1   2   3

> c./c

ans =
    1    1    1   NaN    1    1    1
```

Since division by zero is not defined, we must be careful with the dot-division operation.

In the first case we have ended up with an `Inf` in the result as we have tried to divide by zero. In the second one the result contains a `NaN` (Not a Number) as we have tried to calculate $0/0$.

As it is the case with the dot-star product, the implementation of the dot-division operation enables us to carry out calculations over entire vectors in one single command. Let us, for instance, tabulate the values of the function

$$y(x) = \frac{\sin x}{x}. \tag{2.27}$$

For simplicity we only show here a calculation with five values:

```
> x=0.1:0.2:1

x =
    0.1000    0.3000    0.5000    0.7000    0.9000

> y=sin(x)./x

y =
    0.9983    0.9850    0.9588    0.9203    0.8703
```

Please note that the example above leaves out the case where $x = 0$, where a naive approach would result in a NaN value as we would be dividing by zero. We recommend taking a look at the techniques described in Chapter 5 and in particular Exercise 6 in that chapter.

To deal with the potential division by zero in Equation (2.27) refer to Exercise 6 in Chapter 5.

2.10   Dot-Power of Vectors, .^

THE DEFINITION OF THE DOT-DIVISION explained in the previous section simplifies the number of operations that have to be carried out. If we require to obtain the power of each of the elements on a vector we can use the .^ operator.

The dot-power operator allows us to raise each element of a vector to the desired power.

For instance, if we want to calculate the square of each element of a vector **u**, we can apply the dot-star product defined in Section 2.7, in other words, u.*u.

```
> u = [10, 5, 0];
> u.*u

ans =
    100    25    0
```

However, a more elegant way of doing this is using the dot-power operator:

```
> u.^2

ans =
    100    25    0
```

Taking the square of each element of a vector.

This operation can also be used with any power. For example, we can take the vector u to the power of 3:

```
> u.^3

ans =
    1000    125    0
```

Taking the cube of each element of a vector.

We can also take negative powers; for example, we can take the vector u to the power of -2:

```
> v.^(-2)

ans =
   0.250000
   0.062500
   0.027778
```

2.11 *Summary*

THIS CHAPTER HAS ENABLED US to see how MATLAB and Octave can be far more powerful than a simple calculator by extending the use of various operators from scalars (numbers) to vectors. A vector is thus an arrangement of numbers either in a column or in a row. We have seen how column and row vectors can be constructed with the aid of square brackets. Also, we have seen the power of the colon notation (:) to define lists of numbers to be used in the construction of vectors.

We have learnt how to extract elements of a vector by referring to the position of each element as in index. We have also seen how arithmetic operations such as addition and substraction can be directly applied to vectors. Furthermore, we have covered further operations on vectors such as the scalar product (∗) and element-wise operations such as the dot-star product (.∗), dot-division (./) and dot-power (.^). We have also seen how the orientation of the vectors can make some of the operations mentioned above possible; in order to deal correctly with some of those operations we have introduced the transposition operator (').

In the next chapter we will see how MATLAB and Octave extend the capabilities of vector operations and manipulations to more general arrays: matrices.

2.12 Exercises

1. Create row vectors that contain the following elements:

   (a) 6, 87, 407, $\sqrt{\pi}$, 1/5, ln(10) and cos(45°)

   (b) $\frac{6}{2+3^2}$, $e^{2.5}$, 76, 98, 341

   (c) 1, 3, 5, 7, 9, 11

   (d) 0, 1, 1, 2, 3, 5, 8, 13, 21

2. Create a column vector v1 with equally spaced elements and where the first element is 32 and the last element is 421. Use v1 to create a row vector.

3. Given the variables $a = 0.432$ and $b = 1.654$, create a column vector with the following elements: $a, a^3, b^{-1}, ab, \sqrt[b]{a}$.

4. Using the colon notation create a 1×10 row vector whose elements are all 83.

5. The following vector is defined in MATLAB and Octave:

   $\text{vec1} = [8, 6, 90, -0.14, 56, 76, 7, -2, 0, 0.82176, 10, -54].$

   Find the following sub-vectors:

   (a) v1 with elements 2 through to 5

   (b) v2 made of elements 1, 4, 5, 6, 7 and 10

   (c) v3 made of elements 11, 5, 8, 1, 2, 3

6. Using vector operators, evaluate the following functions for the interval $[-3, 3]$ for 30 equally spaced points:

   (a) $y = -x^3 + 2x - 8$

   (b) $y = \cos(3x) - 4x^2$

   (c) $y = 10x^{(2/3)} + \frac{1}{x+1}$

7. Consider the following two vectors:

$$\mathbf{u} = 9\hat{x} - 7.5\hat{y} + 4.1\hat{z},$$
$$\mathbf{v} = -8.5\hat{x} - 0.3\hat{y} - 7\hat{z}.$$

Calculate the dot product $\mathbf{u} \cdot \mathbf{v}$ in the following ways:

(a) Using vector multiplication

(b) Using the colon notation and the built-in function sum (find out what the built-in function sum does)

(c) Using the built-in function dot (find out what the built-in function dot does)

8. Using MATLAB and Octave corroborate empirically that the infinite series below converges.

$$6 \sum_{n=1}^{\infty} \left(\frac{1}{n^2} \right) = \pi^2.$$

Use $n = 10$, $n = 200$, $n = 1000$, $n = 10000$.

9. Consider the following column vectors:

$$\mathbf{a} = \begin{pmatrix} 9 \\ 5 \\ -1 \\ 3 \end{pmatrix}, \quad \mathbf{b} = \begin{pmatrix} 4 \\ -6.5 \\ 8 \\ 7 \end{pmatrix}.$$

(a) Raise each element of $\mathbf{a}$ to the power of the corresponding element in $\mathbf{b}$.

(b) Divide each element of $\mathbf{a}$ by the corresponding element of $\mathbf{b}$.

(c) Multiply the two vectors element by element.

10. Using the following vectors, $\mathbf{a} = (9, 5, -3)$, $\mathbf{b} = (-6, 0.5, -2)$ and $\mathbf{c} = (8, 4, -2)$, verify that the fol-

lowing vector identity holds:

$$\mathbf{a} \times (\mathbf{b} \times \mathbf{c}) = \mathbf{b}(\mathbf{a} \cdot \mathbf{c}) - \mathbf{c}(\mathbf{a} \cdot \mathbf{b})$$

Use the built-in function `cross` to calculate the cross-product.

3

Matrices and Matrix Operators

IN THE PREVIOUS CHAPTER WE saw how to define vectors in MATLAB® and Octave as well as the way in which we can manipulate and carry out operations with them. As it turns out, row and column vectors are special cases of matrices. An $m \times n$ matrix is a rectangular array of numbers having m rows and n columns:

$$\mathbf{A} = \begin{pmatrix} a_{1,1} & a_{1,2} & \cdots & a_{1,n} \\ a_{2,1} & a_{2,2} & \cdots & a_{2,n} \\ \vdots & \vdots & \ddots & \vdots \\ a_{m,1} & a_{m,2} & \cdots & a_{m,n} \end{pmatrix}. \tag{3.1}$$

A matrix can be thought of as a collection of row (or column) vectors.

We can refer to m and n as the size of the matrix.

Let us consider the following 3×2 matrix:

$$\mathbf{A} = \begin{pmatrix} 12 & 30 \\ 42 & 15 \\ -3 & -7 \end{pmatrix}. \tag{3.2}$$

We can enter this matrix into MATLAB and Octave one row at a time and each row separated by a new line:

```
> A=[12 30
     42 15
     -3 -7]


A =

    12    30

    42    15

    -3    -7
```

A matrix can be entered by separating each row with a new line.

The same syntax that was used for vectors can be applied here: instead of entering a new line to define a row, we can simply separate the rows with semicolons.

```
> B=[9 8; 7 6; 5 4]


B =

    9    8

    7    6

    5    4


> C=[1:3;4:6;7:9]


C =

    1    2    3

    4    5    6

    7    8    9
```

Alternatively we can use semicolons to separate each row.

In this way, the matrices defined above, namely, A and B, are 3×2 matrices whereas C is 3×3. With this notation, we can think of a row-vector as a $1 \times n$ array and a column vector is an $m \times 1$ matrix.

A row-vector can be thought of as a $1 \times n$ array and a column vector an $m \times 1$ one.

3.1   *Size of a Matrix*

THE SIZE OR DIMENSIONS OF a matrix defined in MATLAB
and Octave can be obtained by using the command `size`.

```
> size(A), ...
    size(B,1), ...
    size(B,2)


ans =
    3   2
ans =
    3
ans =
    2
```

The command `size` returns the dimensions of a matrix. The first number corresponds to the number of rows and the second one to the number of columns.

In the example above we can verify the dimensions of the
matrices that we defined in the previous section. Notice
that the first number returned corresponds to the number
of rows, whereas the second one to the number of columns.
We can pass this information to `size` command to obtain the
size along the desired dimension as shown above.

It is also possible to assign the results from the `size` command to an array of variables:

```
> [rows cols]=size(C)


rows =
    3
cols =
    3
```

3.2   *Transpose of a Matrix*

TRANSPOSING A VECTOR CHANGES IT from a row to a column vector and vice versa as we explained in Section 2.5. It is natural to extend the operation to matrices, which results in the interchange of rows and columns. The operation can be carried out in exactly the same way, i.e. by using an apostrophe or single quote (').

The transpose of a matrix can be obtained with the apostrophe or single quote operator (').

```
> D=[1:5;6:10;11:15]/2

D =

   0.50000   1.00000   1.50000   2.00000   2.50000
   3.00000   3.50000   4.00000   4.50000   5.00000
   5.50000   6.00000   6.50000   7.00000   7.50000

> D'

ans =

   0.50000   3.00000   5.50000
   1.00000   3.50000   6.00000
   1.50000   4.00000   6.50000
   2.00000   4.50000   7.00000
```

We obtain the transpose of the matrix D using D'.

Let us see what the result of the command size is for the matrices above:

```
> size(D), size(D')

ans =

   3   5
ans =

   5   3
```

The transposition has swapped rows and columns in the matrix.

An important aspect to consider when using the apostrophe or single quote notation to calculate the transpose of an array is the fact that when applied to a matrix with complex elements, the complex conjugate transpose is actually calculated. For example, consider the matrix

The complex conjugate transpose of a matrix can be obtained with the apostrophe notation.

$$\mathbf{E} = \begin{pmatrix} 1+i & 2+3i & 4-5i \\ 2-5i & 6-6i & 7+i \\ 3+8i & 9-10i & 5+4i \end{pmatrix} \qquad (3.3)$$

whose complex conjugate transpose is given by

$$\mathbf{E}^* = \begin{pmatrix} 1-i & 2+5i & 3-8i \\ 2-3i & 6+6i & 9+10i \\ 4+5i & 7-i & 5-4i \end{pmatrix}. \qquad (3.4)$$

We can obtain the latter in MATLAB and Octave as follows:

```
> E=[1+1i 2+3*1i 4-5*1i; ...
    2-5*1i 6-6*1i 7+1i; ...
    3+8*1i 9-10*1i 5+4*1i]

E =
    1 +  1i    2 +  3i    4 -  5i
    2 -  5i    6 -  6i    7 +  1i
    3 +  8i    9 - 10i    5 +  4i


> E_star = E'

E_star =
    1 -  1i    2 +  5i    3 -  8i
    2 -  3i    6 +  6i    9 + 10i
    4 +  5i    7 -  1i    5 -  4i
```

We can obtain the direct transpose with the transpose command as follows:

```
> E_trans=transpose(E)

E_trans =
    1 +  1i    2 -  5i    3 +  8i
    2 +  3i    6 -  6i    9 - 10i
    4 -  5i    7 +  1i    5 +  4i
```

We can obtain the transpose directly with the transpose command.

3.3 Special Matrices

IN THE SAME WAY THAT special operations were defined for vectors in Section 2.6, it is useful to have at hand matrices with particular entries which turn out to be frequently employed when using MATLAB and Octave. One example is a matrix whose entries are all ones. This can be obtained using the command ones(m,n).

The command ones generates a matrix whose entries are all ones.

Let us for instance create a matrix with three rows and four columns such that its elements are all the number 1:

```
> P1=ones(3,4)

P1 =
    1   1   1   1
    1   1   1   1
    1   1   1   1
```

Using this command to generate the matrix is much faster than specifying each digit 1 using the syntax we have used for defining a matrix.

Another example of a special matrix is that whose elements are all zero. This can be generated with the command zeros(m,n). We can for example create a 4×3 matrix whose entries are all 0:

The command zeros generates a matrix whose entries are all zero.

```
> Z1=zeros(4,3)

Z1 =
     0    0    0
     0    0    0
     0    0    0
     0    0    0
```

We can use the commands above together with the size of a given matrix to generate a new matrix with the same dimensions. Let us have a look at an example by creating a couple of matrices with the same dimensions as the matrix D defined in Section 3.2 whose dimensions were 3×5. In order to do this we can combine size, that returns the dimensions of the matrix, with the special matrices defined above:

```
> ones(size(D)), zeros(size(D))

ans =
     1    1    1    1    1
     1    1    1    1    1
     1    1    1    1    1
ans =
     0    0    0    0    0
     0    0    0    0    0
     0    0    0    0    0
```

We have combined the command size and the special matrices ones and zeros to create two matrices with the same dimensions of matrix as D defined above.

3.3.1   *Square Matrices*

AS WE HAVE SEEN, A matrix is an array of elements arranged in rows and columns. When the number of rows and columns are the same, i.e. an $n \times n$ array, the matrix is called a square matrix.

A square matrix has the same number of rows and columns.

A matrix is said to be symmetric if transposition leaves the matrix unchanged. This can only be the case if the array is a square matrix.

Let us take a look at an example by defining a matrix sym with 4 rows and 4 columns, i.e. a 4×4 matrix:

```
> sym=[1 3 5 7; 3 2 4 8; 5 4 3 9; 7 8 9 4]

sym =
    1    3    5    7
    3    2    4    8
    5    4    3    9
    7    8    9    4
```

Let us now obtain the transpose of this matrix:

```
> sym_transp=sym'

sym_transp =
    1    3    5    7
    3    2    4    8
    5    4    3    9
    7    8    9    4
```

A square matrix is symmetric if its transpose is the same as the original matrix.

If we inspect the second matrix, we can see that the elements are the same as those in the first one; furthermore, we could see that the upper triangle in the matrix is reproduced in the lower triangle. If we wanted to corroborate the fact that the elements are the same, we could take the difference of these two matrices:

```
> sym-sym_transp

ans =
    0   0   0   0
    0   0   0   0
    0   0   0   0
    0   0   0   0
```

As we can see the result is a matrix of zeros; this means that the elements of both matrices are the same and thus the matrix sym is a symmetric matrix.

3.3.2   The Identity Matrix

THE IDENTITY MATRIX IS A square matrix whose elements along the main diagonal are all one and the rest are all zero. This matrix can be obtained by using the command eye(n) to obtain the $n \times n$ identity matrix. Let us generate a 5×5 identity matrix:

The command eye generates the identity matrix.

```
I=eye(5)

I =
    1   0   0   0   0
    0   1   0   0   0
    0   0   1   0   0
    0   0   0   1   0
    0   0   0   0   1
```

Matrices of this type are called *identity* because when a
matrix of the appropriate dimensions is multiplied by it, the
result is the original matrix itself. We will talk more about
matrix multiplication in Section 3.8.

Multiplication of a matrix by the
identity matrix returns the original
matrix.

For now let us take a look at an example. Using the 5 ×
5 identity matrix above we can calculate the following
multiplication:

$$\mathbf{I} \times \mathbf{A} = \begin{pmatrix} 1 & 0 & 0 & 0 & 0 \\ 0 & 1 & 0 & 0 & 0 \\ 0 & 0 & 1 & 0 & 0 \\ 0 & 0 & 0 & 1 & 0 \\ 0 & 0 & 0 & 0 & 1 \end{pmatrix} \begin{pmatrix} 1 \\ 2 \\ 3 \\ 4 \\ 5 \end{pmatrix}. \qquad (3.5)$$

Let us then start by defining the matrix **A** which is given
by a column vector whose elements are the sequence of
numbers from 1 to 5. We can thus use the colon notation to
facilitate its construction:

```
> A=[1:5]'

A =
    1
    2
    3
    4
    5
```

We can now carry out the matrix multiplication (using the *
operator):

We will talk more about matrix
multiplication in Section 3.8.

```
> I*A

ans =
   1
   2
   3
   4
   5
```

As we can see the original matrix **A** was re-obtained.

3.4   Diagonal Matrices

A DIAGONAL MATRIX HAS A very similar structure to the identity matrix defined above. The difference is that the elements along the diagonal are not necessarily all ones:

A diagonal matrix has non-zero elements in the main diagonal.

$$\mathbf{D} = \begin{pmatrix} a_{1,1} & 0 & \cdots & 0 \\ 0 & a_{2,2} & \cdots & 0 \\ \vdots & \vdots & \ddots & \vdots \\ 0 & 0 & \cdots & a_{m,n} \end{pmatrix}. \qquad (3.6)$$

For example, if we want to enter the following diagonal matrix into MATLAB and Octave,

$$D = \begin{pmatrix} 90 & 0 & 0 \\ 0 & 45 & 0 \\ 0 & 0 & 30 \end{pmatrix}, \qquad (3.7)$$

we can certainly do so with the syntax we already know; this can be done as follows:

```
> D = [90 0 0; 0 45 0; 0 0 30]

D =

    90    0    0
     0   45    0
     0    0   30
```

This may work fine for a small matrix as the one shown, but as the number of elements increases the method is not very practical. This can be simplified by using the command diag. The use of this function requires only to define a vector whose values are the entries of the diagonal of the required matrix:

The command diag allows us to define diagonal matrices by specifying a vector whose elements are those in the diagonal.

```
> d=[90, 45 30]

d =

    90   45   30
```

This information can be used to construct the matrix as follows:

```
> D=diag(d)

D =

    90    0    0
     0   45    0
     0    0   30
```

Another important use of this function is the extraction of the diagonal elements of any matrix. This command does not require the matrix to be square. Let us take the following matrix:

The command diag can also be used to extract the diagonal of a matrix.

```
> sym=[1 3 5 7; 3 2 4 8; 5 4 3 9; 7 8 9 4]

sym =
    1   3   5   7
    3   2   4   8
    5   4   3   9
    7   8   9   4
```

We can easily see that the diagonal elements are $1, 2, 3, 4$. These elements can be extracted with the help of the diag command as follows:

```
> diag(sym)

ans =
    1
    2
    3
    4
```

Another important use of the diag command is the manipulation of off-diagonal elements. Let us consider constructing the following matrix:

$$D_1 = \begin{pmatrix} 0 & 1 & 0 & 0 & 0 \\ 0 & 0 & 2 & 0 & 0 \\ 0 & 0 & 0 & 3 & 0 \\ 0 & 0 & 0 & 0 & 4 \\ 0 & 0 & 0 & 0 & 0 \end{pmatrix}, \qquad (3.8)$$

which can easily be created with the use of the diag(v,k) command, whose first argument v is a vector containing the elements in the diagonal, and the second argument k corresponds to the kth diagonal of the matrix, with $k = 0$

diag(v,k) enables us to construct matrices with off-diagonal elements.

being the main diagonal, positive values of k above the main
diagonal and negative values below it. In this case we can
use the following commands to construct the desired matrix:

```
> d = [1:4];
> D1 = diag(d,1)

D1 =
     0   1   0   0   0
     0   0   2   0   0
     0   0   0   3   0
     0   0   0   0   4
     0   0   0   0   0
```

3.5   Building Matrices

IN THE SAME WAY THAT we used parts of vectors to construct
larger ones in Section 2.3, it is possible to construct larger
matrices out of existing smaller ones, rather than writing the
entire array from scratch. This can be explained much more
easily with an example.

Let us consider the following two arrays:

$$T_1 = \begin{pmatrix} 9 & 8 \\ 7 & 6 \\ 5 & 4 \end{pmatrix}, \tag{3.9}$$

$$T_2 = \begin{pmatrix} 3 & 2 & 1 \end{pmatrix}. \tag{3.10}$$

The first one is a 3×2 matrix, whereas the second one is a
row-vector with three elements. Let us create these arrays in
MATLAB and Octave:

```
> T1=[9 8; 7 6; 5 4], T2=[3 2 1]

T1 =
     9    8
     7    6
     5    4


T2 =
     3    2    1
```

We can now use these two matrices to create new ones by concatenating the arrays. For example, we can build a 3×3 matrix by appending the row-vector T_2 to T_1 as a new column. We would need to transpose T_2 and concatenate the two arrays:

Concatenation of arrays can be used to construct new matrices.

```
> T3=[T1 T2']

T3 =
     9    8    3
     7    6    2
     5    4    1
```

In this case there is no need to add a comma between the arrays as we are attaching a column.

Similarly, we could have transposed T_1 and created a new matrix as follows:

```
> T4=[T1'; T2]

T4 =
     9    7    5
     8    6    4
     3    2    1
```

In this case we need to separate the arrays with a semicolon as we are creating new rows.

This method works for larger arrangements. Consider for example the following matrix:

$$\mathbf{p} = \begin{pmatrix} 1 & 2 & 3 \\ 4 & 6 & 8 \\ 12 & 13 & 144 \end{pmatrix}. \qquad (3.11)$$

From this matrix we can think of constructing the following one:

$$\mathbf{p_{new}} = \begin{pmatrix} 1 & 0 & 0 & 1 & 2 & 3 \\ 0 & 2 & 0 & 4 & 6 & 8 \\ 0 & 0 & 3 & 12 & 13 & 14 \\ 0 & 0 & 0 & 1 & 4 & 12 \\ 0 & 0 & 0 & 2 & 6 & 13 \\ 0 & 0 & 0 & 3 & 8 & 14 \end{pmatrix}. \qquad (3.12)$$

This can be achieved by manipulating the elements of the **p** matrix with the commands we have seen so far as follows:

```
> p=[1:3; 4:2:9; 12 13 14]

p =

     1    2    3
     4    6    8
    12   13   14

> p_new=[diag(1:3) p; zeros(3) p']

p_new =

     1    0    0    1    2    3
     0    2    0    4    6    8
     0    0    3   12   13   14
     0    0    0    1    4   12
     0    0    0    2    6   13
     0    0    0    3    8   14
```

The concatenation of matrices can be used to construct larger arrays, too.

IT IS SOMETIMES USEFUL TO visualise the non-zero entries of
a matrix, particularly those with a large number of elements.
We can obtain a visualisation in a Cartesian plane by applying the command spy. In the case above, if we issue the
command spy(p_new) MATLAB and Octave will generate a
plot similar to that shown in Figure 3.1.

The built-in function spy lets us
visualise structure of a matrix.

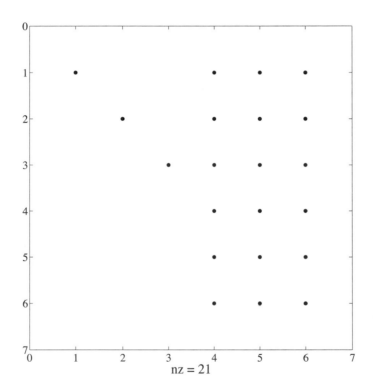

Figure 3.1: Non-zero elements
of the matrix p_new defined
in the text. The command spy
can be used to obtain a visual
representation of the non-zero
elements of a matrix.

3.6 Tabulating Functions

WE ARE NOW FAMILIAR WITH the way MATLAB and Octave
deal with arrays of numbers in columns and rows. We can

exploit this feature of the software to create tables of values obtained from the evaluation of a mathematical function.

Let us consider for example the tabulation of the following two functions in the interval $[0, \pi]$:

$$y_1 = 42\cos(2x), \tag{3.13}$$

$$y_2 = 2\sin(42x). \tag{3.14}$$

This task has been addressed in earlier sections but we are now in a position to produce a more suitable table format.

We start by defining a vector that spans the interval given and can then use it to calculate the value of the desired functions:

```
> x=0:0.5:pi;
> y1=42*cos(2*x);
> y2=2*sin(42*x);
> [x' y1' y2']

ans =
      0.00000   42.00000    0.00000
      0.50000   22.69270    1.67331
      1.00000  -17.47817   -1.83304
      1.50000  -41.57968    0.33471
      2.00000  -27.45303    1.46638
      2.50000   11.91381   -1.94107
      3.00000   40.32715    0.65998
```

The effective usage of the commands we have seen (vectorisation, concatenation, transposition, etc.) provides us with an easy way to tabulate functions.

We have chosen to create a vector x ranging from 0 to π with points located every 0.5 units. This vector was used to construct two more vectors (y_1 and y_2) and finally we output the result by concatenating the transposed row-vectors.

We can increase the number of values by changing the step in the first command. A similar result could have been obtained with the command [x; y; u;]' or by transposing the vector x at the beginning, i.e. x=(0:0.5:pi)'.

Remember that there are multiple ways to obtain the same result.

3.7   Extracting Parts of Matrices

THE WAY IN WHICH MATLAB and Octave treat vectors can readily be applied to matrices, for example in the extraction of elements from matrices. The discussion here follows the one presented in Section 2.3.

In order to extract parts of matrices we have to take into account that each and every element of a matrix is indexed according to the row and column where it is positioned. In this way, $A_{(r,c)}$ refers to the element in the rth row and c^{th} column of the matrix **A**. In MATLAB and Octave this is denoted by A(r,c).

Element extraction relies on the indexation of the elements in terms of rows and columns.

Let us consider an example: Given the matrix

$$\mathbf{A} = \begin{pmatrix} 16 & 2 & 3 & 13 \\ 5 & 11 & 10 & 8 \\ 9 & 7 & 6 & 12 \\ 4 & 14 & 15 & 1 \end{pmatrix}, \qquad (3.15)$$

we can see that the element in the first row and first column is $A_{(1,1)} = 16$, the element in row 4, column 3 is $A_{(4,3)} = 15$ and the element in row 3, column 1 is $A_{(3,1)} = 9$.

This can also be achieved in MATLAB and Octave. First, we note that the matrix A above is in fact a 4 × 4 magic matrix, and as such we can easily create it with the use of the magic command:

```
> A=magic(4)

A =

    16     2     3    13
     5    11    10     8
     9     7     6    12
     4    14    15     1
```

We can now obtain the elements mentioned above by referring to them using the indices in the matrix:

```
> A(1,1)

ans =

    16

> A(4,3)

ans =

    15

> A(3,1)

ans =

     9
```

We can refer to elements in a matrix A with the notation A(r,c), where r refers to the row and c to the column.

If we refer to rows and/or columns that are out of bounds for the definition of the matrix used, the software will return an error:

```
> A(5,1)

Index exceeds matrix dimensions.
```

MATLAB

```
> A(5,1)


error: A(I,J): row index out of bounds;
    value 5 out of bound 4
```

Octave

We can also carry out operations with individual elements of the matrix. For example, let us imagine that we require the element $A_{(1,1)}$ to be equal to the element $A_{(4,4)} + 9$; we can do this by operating on the element alone as follows:

```
> A(1,1) = A(4,4) + 9;
```

Note that the operation above only affected that particular element in the matrix. We can corroborate this by displaying the full matrix **A**:

```
> A

A =

   10    2    3   13
    5   11   10    8
    9    7    6   12
    4   14   15    1
```

Let us see another example. We want the element $A_{(4,4)}$ to be three times $A_{(2,2)}$ and subtract the value of $A_{(1,1)}$.

```
> A(4,4)=3*A(2,2)-A(1,1)


A =
   10    2    3   13
    5   11   10    8
    9    7    6   12
    4   14   15   23
```

The operation on individual matrix elements only affects those elements alone.

Again, the operation only affected the chosen element and nothing else.

WE CAN ALSO USE THE colon notation to extract entire rows and columns. For example, given the 5×5 magic matrix:

$$\mathbf{B} = \begin{pmatrix} 17 & 24 & 1 & 8 & 15 \\ 23 & 5 & 7 & 14 & 16 \\ 4 & 6 & 13 & 20 & 22 \\ 10 & 12 & 19 & 21 & 3 \\ 11 & 18 & 25 & 2 & 9 \end{pmatrix}, \qquad (3.16)$$

we are interested in extracting the second column. This can be achieved in MATLAB and Octave as follows:

```
> B=magic(5)

B =

    17    24     1     8    15
    23     5     7    14    16
     4     6    13    20    22
    10    12    19    21     3
    11    18    25     2     9

> B(:,2)

ans =
    24
     5
     6
    12
    18
```

The colon notation can be used with matrices to refer to a sequence of elements in a matrix.

Here we have used the colon notation to refer to all the rows in the first index, and specifying column number 2. If we are now interested in extracting the third and fourth rows, we can do the following:

```
> B(3:4,:)

ans =
    4    6   13   20   22
   10   12   19   21    3
```

We can also obtain a submatrix; for instance, if we are interested in the 2 × 2 submatrix given by the elements from the third and forth rows and columns we can do the following:

```
> B(3:4,3:4)

ans =
   13   20
   19   21
```

It should be clear that using the colon operator, :, on its own refers to either entire rows or columns; for example, B(:,:) is exactly the same as simply B.

Using B(:,:) is equivalent to simply referring to the matrix B directly.

3.8   Matrix Multiplication

MATRIX MULTIPLICATION IS EFFECTIVELY A binary operation between a pair of arrays and the result of that operation is another array. MATLAB and Octave distinguish between a true matrix multiplication as defined in the mathematical sense and an element-by-element multiplication of arrays. Let us take a look at them.

MATLAB and Octave distinguish between true matrix operation and element-by-element multiplication.

3.8.1 Dot-Star Product of Matrices, .∗

IN SECTION 2.8 WE HAVE seen how MATLAB and Octave treat the multiplication of two vectors with the .∗ notation: the multiplication is carried out on an element-by-element basis. In a similar way, in the dot-star product of matrices corresponding elements are multiplied together. This means that the matrices involved must have the same dimensions.

The .∗ operator is used to carry out element-by-element matrix multiplications.

Let us consider the following example with the matrices defined below:

$$\mathbf{M_1} = \begin{pmatrix} 1 & 2 & 3 & 4 \\ 5 & 6 & 7 & 8 \end{pmatrix}, \tag{3.17}$$

$$\mathbf{M_2} = \begin{pmatrix} 11 & 12 & 13 & 14 \\ 15 & 16 & 17 & 18 \end{pmatrix}. \tag{3.18}$$

The element-wise multiplication would be given by

$$\mathbf{M_3} = \mathbf{M_1}.\ast\mathbf{M_2} = \begin{pmatrix} 1 \times 11 & 2 \times 12 & 3 \times 13 & 4 \times 14 \\ 5 \times 15 & 6 \times 16 & 7 \times 17 & 8 \times 18 \end{pmatrix},$$

$$= \begin{pmatrix} 11 & 24 & 39 & 56 \\ 75 & 96 & 119 & 144 \end{pmatrix}. \tag{3.19}$$

This can be carried out with MATLAB and Octave as follows: For $\mathbf{M_1}$ we have

```
> M1=[1:4; 5:8]

M1 =
     1   2   3   4
     5   6   7   8
```

whereas for $\mathbf{M_2}$

```
M2=[11:14;15:18]

M2 =
    11    12    13    14
    15    16    17    18
```

and thus the dot-start multiplication is given by

```
> M3=M1.*M2
M3 =
    11    24    39    56
    75    96   119   144
```

As we can see the element $M_{3(r,c)} = M_{1(r,c)}M_{2(r,c)}$, where r and c are the indices of the rows and columns for each element in the matrix.

From the example above it must be clear that if the dimensions of the matrices do not match, the dot product will throw an error. For example, if we were to calculate

$$\mathbf{M_1}. * \mathbf{M_2^T},$$

the dimensions of the matrices do not match and we would obtain an error:

The dimensions of the matrices must match for the dot-star product to be properly defined.

```
> M4=M1.*M2'

Error using  .*
Matrix dimensions must agree.
```

MATLAB

```
> M4=M1.*M2'

error: product: nonconformant arguments
    (op1 is 2x4, op2 is 4x2)
```

Octave

In the examples above we have transposed the matrix M_2 which makes the product undefined.

3.8.2   Matrix-Vector Products

A VECTOR IS AN ARRAY whose elements are arranged either in a row or in a column. We can therefore define the product of a matrix with a vector and it must be clear that this product can only be defined if a column vector has the same number of elements as the matrix has columns.

In other words, if the matrix A is an $m \times n$ matrix and x is a column vector of length n, then the matrix-vector product Ax can indeed be performed. In this manner, an $m \times n$ matrix times an $n \times 1$ matrix results in an $m \times 1$ matrix. For example, given the following arrays:

Given appropriate dimensions, the product of a matrix and a vector is well defined.

$$A = \begin{pmatrix} 7 & 1 & 4 \\ 11 & 79 & 42 \\ 10 & 15 & 20 \end{pmatrix}, \tag{3.20}$$

$$x = \begin{pmatrix} 32 \\ 15 \\ 9 \end{pmatrix}, \tag{3.21}$$

we can carry out the matrix multiplication as follows:

$$
\mathbf{Ax} = \begin{pmatrix} 7 & 1 & 4 \\ 11 & 79 & 42 \\ 10 & 15 & 20 \end{pmatrix} \begin{pmatrix} 32 \\ 15 \\ 9 \end{pmatrix}, \qquad (3.22)
$$

$$
= \begin{pmatrix} 7 \times 32 + 1 \times 15 + 4 \times 9 \\ 11 \times 32 + 79 \times 15 + 42 \times 9 \\ 10 \times 32 + 15 \times 15 + 20 \times 9 \end{pmatrix}. \qquad (3.23)
$$

The product above can be carried out in MATLAB and Octave as follows:

```
> A=[7, 1, 4; 11, 79, 42; 10 15 20]

A =

     7     1     4
    11    79    42
    10    15    20

> x=[32; 15; 9]

x =

    32
    15
     9

> A*x

ans =
   275
  1915
   725
```

The * operator is used to carry out matrix-vector products of appropriate dimensions.

Let us emphasise that the matrix-vector product is not commutative and in some cases not only would the result be different, but it may not even be defined, as we can see in the following example:

```
> x*A
Error using  *
Inner matrix dimensions must agree.
```

MATLAB

```
> x*A
error: operator *:
  nonconformant arguments (op1 is 3x1, op2 is 3x3)
```

Octave

We obtained this error because the multiplication of the vector **x** with the matrix **A** does not fulfil the condition stated above about the dimensions of the arrays involved.

If the dimensions of the arrays do not match the software will throw an error.

3.8.3   *Matrix-Matrix Products*

WE CAN THINK OF A matrix as a group of column vectors placed next to each other. In that way, the definition of a matrix-matrix product is an extension of the matrix-vector multiplication we have seen in the previous section. The product of an $m \times n$ matrix **A** and an $n \times p$ matrix **B**, written as **AB**, results in $m \times p$ matrix. The number of columns of matrix **A** must match the number of rows of matrix **B**.

The matrix-matrix product is carried out in the same way as the matrix-vector one.

Let us look at an example: with the matrix **A** defined above in Expression (3.20),

$$\mathbf{A} = \begin{pmatrix} 7 & 1 & 4 \\ 11 & 79 & 42 \\ 10 & 15 & 20 \end{pmatrix},$$

and a new matrix **B** given by

$$\mathbf{B} = \begin{pmatrix} 1 & 4 \\ 2 & 5 \\ 3 & 6 \end{pmatrix},$$
(3.24)

we can carry out the matrix multiplication **AB**:

$$\mathbf{C} = \mathbf{AB} = \begin{pmatrix} 7 & 1 & 4 \\ 11 & 79 & 42 \\ 10 & 15 & 20 \end{pmatrix} \begin{pmatrix} 1 & 4 \\ 2 & 5 \\ 3 & 6 \end{pmatrix},$$

$$= \begin{pmatrix} 21 & 57 \\ 295 & 691 \\ 100 & 235 \end{pmatrix},$$
(3.25)

where we have left out the actual calculation and given the result directly. This operation can be done in the software by creating matrix **A**:

```
> A=[7, 1, 4; 11, 79, 42; 10 15 20]

A =

      7    1    4
     11   79   42
     10   15   20
```

Similarly we can enter matrix **B**:

```
> B=[1, 4; 2, 5; 3, 6]

B =

     1   4
     2   5
     3   6
```

and thus the matrix-matrix multiplication can be performed
with the * operator:

```
> C=A*B


C =

      21     57
     295    691
     100    235
```

The * operator is able to perform
matrix-matrix multiplications with
appropriate dimensions.

The matrix **C** above is the result of a 3×3 by a 3×2 matrix,
and thus it is a 3×2 array. If we try to obtain the multiplication of **BA** we will obtain an error as the dimensions do
not match. However, if we transpose B the product can be
carried out:

```
> D=B'*A


D =

      59    204    148
     143    489    346
```

Once again, care must be taken
with matching the dimensions of
the matrices to be multiplied.

3.9   *Sparse Matrices*

IN A NUMBER OF APPLICATIONS in physics, engineering,
finance, etc., it is quite common to deal with big matrices
with a large number of zero elements. These type of matrices are called *sparse matrices*. The computational expense
when carrying out operations with these kind of matrices
can be quite high and that is why MATLAB and Octave
include a number of techniques to deal with them in a more
effective manner.

Let us consider a 6×6 matrix **S** with only four non-zero values: $S_{(1,1)} = 1$, $S_{(3,4)} = 3$, $S_{(4,5)} = 42$ and $S_{(6,1)} = 7$, and therefore the 32 other elements are zero. Let us take a look at this matrix:

$$\mathbf{S} = \begin{pmatrix} 1 & 0 & 0 & 0 & 0 \\ 0 & 0 & 0 & 0 & 0 \\ 0 & 0 & 0 & 3 & 0 \\ 0 & 0 & 0 & 0 & 42 \\ 0 & 0 & 0 & 0 & 0 \\ 7 & 0 & 0 & 0 & 0 \end{pmatrix}. \qquad (3.26)$$

The matrix shown has a large number of elements equal to zero. This can be better handled as a sparse matrix.

If we wanted to enter this matrix in the software we can indeed list every single element (including the 32 zeros), but this is not a very effective way of entering such a matrix and larger examples can get too cumbersome. Alternatively, we can define a matrix with the command `zeros` and modify the non-zero entries individually. With larger arrays we also have the added problem of available memory and in these cases it is better to define a sparse matrix, i.e. a matrix where only the non-zero values are specified.

One easy way to define the matrix used in this example is to define three vectors:

1. A vector that contains the row indices of the non-zero elements

2. A vector that contains the column indices of the non-zero elements

3. A vector that holds the values of each of the non-zero elements of the matrix

With this information, we can now use the command `sparse` to build the desired matrix. We can define the row indices:

```
> r=[1,3,4,6]

r =

    1   3   4   6
```

We define a vector with the row indices of the non-zero elements of the sparse matrix.

as well as the column indices:

```
> c=[1,4,5,1]

c =

    1   4   5   1
```

We also need a vector with the column indices of the non-zero elements of the sparse matrix.

and finally the elements themselves:

```
> v=[1, 3, 42, 7]

v =

    1   3   42   7
```

Finally, we need the non-zero elements themselves.

With this information we can use the sparse command to generate the matrix by passing the row-index and column-index vectors and the values of the elements:

The sparse command generates sparse matrices.

```
> S=sparse(r,c,v)

S =
    (1,1)        1
    (6,1)        7
    (3,4)        3
    (4,5)       42
```

MATLAB

```
> S=sparse(r,c,v)


S =

  Compressed Column Sparse ...
  (rows = 6, cols = 5, nnz = 4 [13%])


  (1, 1) ->  1
  (6, 1) ->  7
  (3, 4) ->  3
  (4, 5) ->  42
```

Octave

Notice that the result only lists the values of the non-zero elements, and the definition of the matrix is more straight-forward than typing every single zero. In order to convince ourselves that we indeed have the desired matrix, we can obtain a full matrix by using the command full:

```
> SFull=full(S)


SFull =
    1    0    0    0    0
    0    0    0    0    0
    0    0    0    3    0
    0    0    0    0    42
    0    0    0    0    0
    7    0    0    0    0
```

The full command allows us to see the full form of a sparse matrix.

Although the information held by the matrix S and Sfull is the same, they have been defined as different types in the software. The former is a sparse matrix that uses less memory, whereas the latter is a full matrix and thus it uses more memory. It is important to mention that sparse matrices can be used and manipulated in the same way as normal matrices and thus it is not necessary to use the `full` command when doing operations with them.

In MATLAB, we can easily see that a matrix is sparse issuing the command `whos` which lets us see the variables that are currently active in the software:

Name	Size	Bytes	Class	Attributes
S	6x5	112	double	sparse
Sfull	6x5	240	double	

MATLAB

In the case of Octave the attributes are unfortunately not displayed (up to version 3.8.0 of the software). It may be possible that this behaviour is implemented in later versions of the software.

LET US TAKE A LOOK at a more complicated example by putting together some instructions that allow us to construct the following tridiagonal matrix, for any given value n:

$$B = \begin{pmatrix} 1 & 2 & & & & \\ 1 & 2 & 2 & & & \\ & 1 & 3 & 2 & & \\ & & \ddots & \ddots & \ddots & \\ & & & 1 & n-1 & 2 \\ & & & & 1 & n \end{pmatrix}, \quad (3.27)$$

A more complicated example of a sparse matrix. Since the matrix has non-zero diagonals we will use the `spdiags` to define it.

where the main diagonal contains the numbers from 1 to n, the off-diagonal below the main diagonal is made out of ones, whereas the off-diagonal just above the main diagonal contains only twos, while the rest of the elements are zero (not shown in Equation (3.27)).

In this case we need to define three column vectors, one for each diagonal of non-zero elements. Once these column vectors are created we can put the desired matrix together using the command for sparse diagonals: spdiags.

Let us call these column vectors l for lower diagonal, d for diagonal and u for upper diagonal. These vectors **must have the same length** but different elements are to be used:

spdiags requires the argument vectors to be of the same length, although only a subset of their elements will be used.

- in the case of l we need the first $n - 1$ elements,

- for u only the last $n - 1$ elements are required, and

- for d the full list of elements will be used.

The spdiags command places the l, d and u vectors in the diagonals labelled $-1, 0, 1$. Notice that the main diagonal is identified with 0, negative values correspond to diagonals below the main one, while positive values correspond to those above it.

Let us construct the matrix for the case $n = 5$. We can start by defining the elements that go in l, d and u:

```
> n=5;

> l=ones(1,n)';  d=(1:n)';  u=2*ones(1,n)';
```

The command spdiags(M,d,m,n) creates an $m \times n$ sparse matrix from the columns of M and places them along the diagonals specified by the array d. In that way, we can now construct our diagonal matrix as follows:

```
> B=spdiags([l d u], -1:1, n, n)

B =
    (1,1)        1
    (2,1)        1
    (1,2)        2
    (2,2)        2
    (3,2)        1
    (2,3)        2
    (3,3)        3
    (4,3)        1
    (3,4)        2
    (4,4)        4
    (5,4)        1
    (4,5)        2
    (5,5)        5
```

MATLAB

spdiags takes the columns from M to create an $m \times n$ matrix, placing the elements according to d.

where we have only shown the output as displayed in MATLAB; the output for Octave is very similar except that the values of the non-zero elements are indicated with ->.

Remember that a sparse matrix only lists its non-zero values when it is displayed. We can see the entire matrix with the full command:

```
> Bfull=full(B)

Bfull =
      1     2     0     0     0
      1     2     2     0     0
      0     1     3     2     0
      0     0     1     4     2
      0     0     0     1     5
```

3.10   Systems of Linear Equations

Now that we know how to enter matrices in MATLAB
and Octave it becomes natural to ask the software to solve
a linear system of equations given that such systems can be
written in terms of matrices.

A general system of linear equations can be expressed in
terms of a coefficient matrix **A**, a column vector **b** and a
column vector of unknowns **x** such that

$$\mathbf{A}\mathbf{x} = \mathbf{b} \qquad (3.28)$$

The use of matrices in solving linear systems becomes quite natural in MATLAB and Octave.

or, expanding the matrix multiplication,

$$
\begin{aligned}
a_{1,1}x_1 + a_{1,2}x_2 + \cdots + a_{1,n}x_n &= b_1, \qquad (3.29)\\
a_{2,1}x_1 + a_{2,2}x_2 + \cdots + a_{2,n}x_n &= b_2,\\
&\vdots\\
a_{n,1}x_1 + a_{n,2}x_2 + \cdots + a_{n,n}x_n &= b_n.
\end{aligned}
$$

When **A** is non-singular and square ($n \times n$), meaning that the
number of independent equations is equal to the number of
unknowns, the system has a unique solution given by

$$\mathbf{x} = \mathbf{A}^{-1}\mathbf{b}, \qquad (3.30)$$

where $\mathbf{A}^{-1}$ is the inverse of the matrix **A**. This means that
the solution vector **x** can be calculated by taking the inverse
of the coefficient matrix **A** and multiplying it from the right
with vector **b**. Although this method seems to be quite
straightforward, in many cases it is not advisable to be

carried out. This is because obtaining the inverse of a matrix is a highly non-trivial task and numerical errors can be introduced. Other methods such as Gaussian elimination are preferred and can be more efficient. Nonetheless, for the purposes of this chapter we shall continue the discussion with the inverse matrix method.

Calculating the inverse of a matrix is not trivial and this method can introduce numerical errors. Be careful!

Let us enter the following square matrix and column vector into the software and solve the linear system $\mathbf{Ax} = \mathbf{b}$:

$$\mathbf{A} = \begin{pmatrix} 5 & 6 & 2 \\ 9 & 7 & -1 \\ 5 & -3 & 2 \end{pmatrix}, \qquad (3.31)$$

$$\mathbf{b} = \begin{pmatrix} 1 \\ 2 \\ 3 \end{pmatrix}. \qquad (3.32)$$

```
> A=[5 6 2; 9 7 -1; 5 -3 2]

A =

        5       6       2
        9       7      -1
        5      -3       2

> b=[1; 2; 3]

b =

        1
        2
        3
```

We can now calculate the inverse of **A** and multiply it to **b** on the right: $x = A^{-1}b$. The inverse of a matrix can be calculated with the command inv.

```
> Ainverse=inv(A)

Ainverse =
   -0.0531     0.0870     0.0966
    0.1111    -0.0000    -0.1111
    0.2995    -0.2174     0.0918

> x=Ainverse*b

x =
    0.4106
   -0.2222
    0.1401
```

The inverse of a matrix can be obtained with inv.

As we mentioned above, this approach based on the inverse of a matrix is correct but it may not be the most efficient way of solving the problem. This is because the number of operations can become very large as well as the procedure is prone to numerical errors unless appropriate techniques are used.

MATLAB and Octave have a number of algorithms and techniques pre-programmed already in their libraries. Moreover, these algorithms can be automatically invoked by the software with a few simple characters. For example, in the case of finding the solution of a linear system of equations, the standard solution routine can be called when using the matrix left-division operator: x = A \ b.

```
> x = A \ b


x =

     0.4106

    -0.2222

     0.1401
```

The left division operator, \, can be used to solve linear systems in MATLAB and Octave.

As we can see we have obtained the same solution in both cases, except that in the second example we did not have to calculate explicitly the inverse matrix. Instead we relied on the software libraries to carry out the operations.

3.11   Summary

IN THIS CHAPTER WE HAVE covered the use of matrices in MATLAB and Octave. We have seen how matrices can be understood as natural extensions of the concept of row and column vectors. A matrix is thus a collection of column vectors next to each other, or alternatively a collection of stacked row vectors. In that sense, the use of the colon notation (:) to define lists or sequences of numbers can still be used here. The elements of a matrix can be extracted and manipulated using the row and column indices in a very straightforward manner.

We also learned how to carry out important operations on matrices, such as the transpose with the use of the apostrophe or single quote (') as well as the command transpose. We also saw how to define special matrices with the aid of commands such as ones and zeros. Important matrices such as the identity matrix and diagonal matrices can be easily constructed with the eye and diag commands.

Similarly, we also saw how to perform common operations on matrices such as addition, substraction and multiplication. In particular we covered two types of multiplication: the dot-star (.*) product (element-by-element) and the star product (*) which corresponds to true matrix multiplication.

Finally, we covered the construction of sparse matrices with the sparse and spdiag commands and used MATLAB and Octave to solve linear systems of equations.

In the next chapter we will see the powerful visualisation features that are included in MATLAB and Octave. They will enable us to analyse and solve problems in a more integrated manner than many other programming systems.

3.12   Exercises

1. Use MATLAB and Octave to enter the following matrix:

$$
M = \begin{pmatrix}
15 & 9 & -i & 8 \\
-9 & 4 & 0.4 & -9.1 \\
1+i & 4 & -7 & 3 \\
4 & 3.3 & 9 & 6.5
\end{pmatrix}.
$$

   (a) Create a vector M_1 with the elements in the second row of M.

   (b) Create a vector M_2 with the elements of the third column of matrix M.

   (c) Create a vector M_3 with the elements in the diagonal of matrix M.

2. Use MATLAB and Octave to enter the following matrix:

$$
M = \begin{pmatrix}
4 & 7 & 9 & 0 \\
8 & -42 & 5 & -\pi \\
4 & 2 & 0 & 1 \\
-6 & -5 & -4 & -3
\end{pmatrix}.
$$

   (a) Create a 3×4 matrix with the elements of the second through to fourth columns of M.

   (b) Create a 2×2 matrix with the elements at the centre of matrix M.

   (c) Create a 2×3 matrix with the elements of the first two columns and last three rows of matrix M.

3. Given the matrices

$$
A = \begin{pmatrix}
10 & 15 \\
8 & 6
\end{pmatrix}, \qquad
B = \begin{pmatrix}
-1 & 5 \\
0.5 & 6.3
\end{pmatrix},
$$

   calculate the following:

(a) $\mathbf{A} + \mathbf{B}$

(b) $\mathbf{B} - \mathbf{A}$

(c) $\mathbf{BA}$

(d) $\mathbf{AB}$

(e) Element-wise multiplication of $\mathbf{A}$ and $\mathbf{B}$

4. Consider the magic square `magic(10)`. Verify that the sum of the elements in each column, each row, the diagonal and antidiagonal add up to the same value. Hint: find out what the `fliplr` function does to deal with the antidiagonal.

5. Find an easy way to generate the following matrices:

(a)

$$\mathbf{M_1} = \begin{pmatrix} 0 & 0 & 0 & 0 \\ 0 & 0 & 0 & 0 \end{pmatrix}$$

(b)

$$\mathbf{M_2} = \begin{pmatrix} 0 & 0 & 1 & 0 & 0 \\ 0 & 0 & 0 & 1 & 0 \\ 0 & 0 & 0 & 0 & 1 \\ 0 & 0 & 0 & 0 & 0 \\ 0 & 0 & 0 & 0 & 0 \end{pmatrix}$$

(c) For the following matrix, use vector multiplication in conjunction with the `ones` command. For an alternative method, explore what the function `repmat` does and make use of it.

$$\mathbf{M_3} = \begin{pmatrix} 1 & 2 & 3 & 4 & 5 \\ 1 & 2 & 3 & 4 & 5 \\ 1 & 2 & 3 & 4 & 5 \\ 1 & 2 & 3 & 4 & 5 \end{pmatrix}$$

6. Tabulate the following functions in the interval $[-10,10]$ in a grid of 100 points:

(a) $y = \ln(x^2) + 1$

(b) $y = \frac{1}{x+7^2}$

(c) $y = 3\exp\left(-x^2\right)$

7. Find out what the functions randn and reshape do. Use one of the functions to create a 10×10 matrix with random elements. Find the maximum values in each column, each row and overall with the help of the max function. Use the other function to transform the 10×10 matrix into a 25×4 matrix.

8. Create a sparse matrix such that

$$\mathbf{A} = \begin{pmatrix} \mathbf{1} & \mathbf{a}^T \\ \mathbf{a} & \mathbf{I}_n \end{pmatrix},$$

where $\mathbf{a}$ is an $n \times n$ matrix, $\mathbf{1}$ is a matrix of ones and $\mathbf{I}_n$ is an $n \times n$ identity matrix. If $\mathbf{a}$ is a matrix with random elements, with $n = 100$, create the corresponding matrix $\mathbf{A}$ and visualise its structure.

9. Consider the matrix

$$\mathbf{B} = \begin{pmatrix} 2 & 1 & -1 \\ 1 & 2 & 1 \\ 1 & 1 & 2 \end{pmatrix}.$$

Using the command eig find the eigenvalues and eigenvectors of $\mathbf{B}$, $\mathbf{B}^T$ and $\mathbf{B}^{-1}$. What relationship (if any) is there between your results.

Do the same for the matrix $\mathbf{B1}$ below. What do the eigenvalues tell you about the matrix?

$$\mathbf{B1} = \begin{pmatrix} 2 & 1 & -1 \\ 1 & 2 & 1 \\ 1 & 1 & 0 \end{pmatrix}$$

10. Using MATLAB and Octave, solve the following linear systems of equations:

(a)

$$
\begin{aligned}
2w - x + 5y + z &= 3 \\
3w + 2x + 2y - 6z &= -32 \\
5w - 2x - 3y + 3x &= 49 \\
w + 3x + 3y - z &= -47
\end{aligned}
$$

(b)

$$
\begin{aligned}
x_1 + x_2 + x_3 &= 4 \\
-3x_1 + 2x_2 - 5x_3 &= -14 \\
2x_1 - 3x_2 + 4x_3 &= 10
\end{aligned}
$$

4
Plotting

WE ARE NOW FAMILIAR WITH some of the most essential elements of MATLAB® and Octave, namely, vectors and matrices. We have learned how to manipulate them and carry out important operations with them such as addition, substraction, multiplication and, within the definition in the software, even division. Similarly, we know how to address and extract individual elements as well as sequences of elements.

In this chapter we turn our attention to one of the distinguishing features of MATLAB and Octave, i.e. the plotting and visualisation capabilities integrated with the development environment itself. Whereas other programming environments do not include a way to produce plots and graphs, MATLAB and Octave enable us to merge data visualisation directly to our workflow.

MATLAB and Octave integrate data visualisation directly to the programming environment.

4.1 *Plotting Simple Functions*

WE HAVE HAD A LOOK at how to enter and evaluate a number of expressions and built-in functions in MATLAB and Octave. In this chapter we are going to see how to visualise those expressions in the form of plots.

Let us consider the function

$$y(x) = \cos(4x), \qquad (4.1)$$

and imagine that we are interested in creating a graph of this function in the interval $0 \le x \le \pi$. This can be easily done by

- taking a sample of points,

- evaluating the function at those points and

- joining the calculated values with appropriate lines.

A common and simple approach is to take equally spaced points along the x values. We can thus have $n + 1$ points a distance h apart from each other; for example, in the case of $n = 20$ we can write:

```
> n = 20;
> h = pi/n;
> x = 0:h:pi;
```

Setting up an equally spaced vector of points.

This will therefore create a set of points stored in the variable x. We could also use the command linspace, whose format is

```
linspace(a,b,n)
```

Equidistant-element vectors can also be created with the linspace command.

which generates a vector with n points at an equal distance from each other between the values a and b. So for the case above we can write

```
> x = linspace(0,pi,21);
```

It is now possible to evaluate *y* at the various points given by the values stored in the vector x:

```
> y = cos(4*x);
```

Finally the plot of the points calculated above can be obtained by using the plot command

```
> plot(x,y)
```

The plot command can be used to produce graphs.

The result of the commands above is shown in Figure 4.1, where it is clear that the number of points used is small given the jagged profile obtained.

The jagged result can easily be improved upon by increasing the number of points to, for instance, 200:

```
> N = 199;
```

Note that we need to specify $N - 1$ so that the colon notation generates a vector with N elements. All we need to do now is repeat the calculations and recreate the plot with the increased number of points:

```
> h = pi/N;
> x = 0:h:pi;
> y = cos(4*x);
> plot(x,y)
```

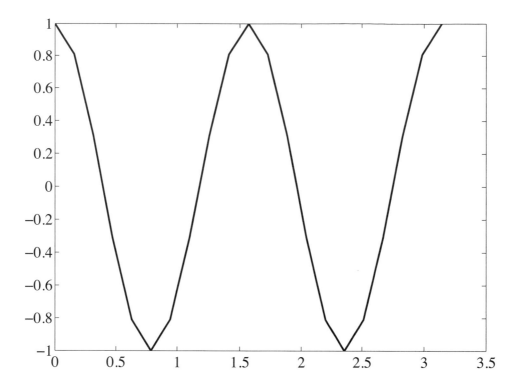

Figure 4.1: Plot of the function $y = \cos(4x)$ generated with 21 sample points. The low sampling generates a jagged profile.

which gives us the plot shown in Figure 4.2.

As we can see, the curve is now much smoother and closer to what we imagine as a plot for the cosine function. The number of points can further be increased, but care must be taken when dealing with larger and more complicated problems, as an increased number of points may result in a large memory use.

Although we have now been able to create a plot for the desired function, we know that appropriate labels and information must be placed in the graph; we will deal with this in the next section.

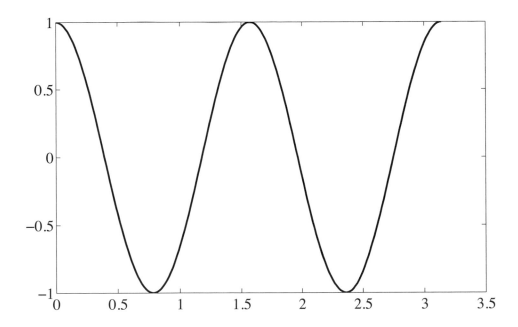

Figure 4.2: Plot of the function $y = \cos(4x)$ generated with 200 sample points. Increasing the number of points gives us a smoother curve.

4.2 Information in the Plot

THE PLOTS SHOWN IN FIGURES 4.1 and 4.2 do show the main characteristics of the function depicted, but they can be made more useful by adding further information such as a helpful and explanatory title, labels for the axes used and perhaps even a legend for the functions plotted. Furthermore, we can increase or decrease the fonts, manipulate the colours used, etc. In this section we will see how this can be done.

4.2.1   *Titles and Labels*

IT IS VERY USEFUL TO add a title that describes the plot as well as information about what it is that is being plotted along each of the axes. In order to add a title and label the axes in the plot, we use the commands `title`, `xlabel` and `ylabel`. Try out the following commands for the plot generated in the previous section; the result can be seen in Figure 4.3.

```
> title('Graph of y = cos(4 x)')
> xlabel('x-axis')
> ylabel('y-axis')
```

The commands `title`, `xlabel` and `ylabel` let us add a title to the plot and label its axes.

The strings enclosed in single quotes can be (almost) anything we choose. Some simple LaTeX commands are available for formatting mathematical expressions and Greek characters. More information about this can be seen in Section 4.6.

The text included in the title and labels supports the use of some LaTeX commands.

4.2.2   *Grids*

SOMETIMES IT IS USEFUL TO show a grid that helps guide the eye when looking at a plot. For example, a dotted grid may be added by issuing the following command:

```
> grid
```

The command `grid` enables a dotted grid to be shown in a plot.

In cases where this is not required, the grid can be removed using either `grid` again, or `grid off`.

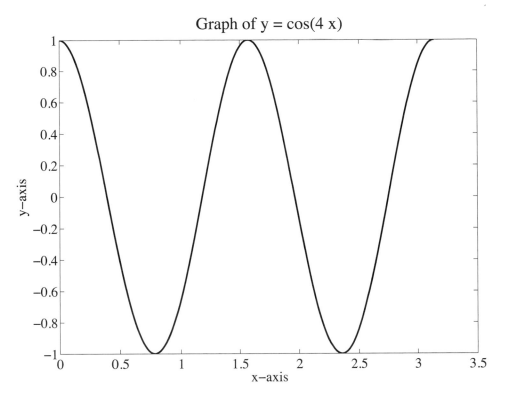

Figure 4.3: Plot of the function $y = \cos(4x)$ including a title and labels for each axis.

4.2.3   Line Styles and Colours

WHEN CREATING A PLOT, THE default is to present the graph with a blue solid line. A more general command to create a plot specifying a solid blue line is as follows:

```
> plot(x,y,'b-')
```

A more general form for the plot command specifies the colour and type of line to be used.

From the discussion in the previous sections we know the meaning of the first two arguments passed to the plot command. This leaves us with the third argument, which is a string (we know this because it is enclosed in single quotation marks): the first character of the string specifies the colour of the line to be plotted, and the second corresponds to the type of line style. In the example above, b stands for blue, and - represents a solid line. The options for colours and styles are shown in Table 4.1. Please note that we can obtain a full list with `help plot`.

Colours		Styles	
b	blue	o	circle
c	cyan	- .	dash dot
g	green	- -	dashed
k	black	:	dotted
m	magenta	+	plus
r	red	.	point
w	white	-	solid
y	yellow	*	star
		x	x mark

Table 4.1: Colours and line styles that can be used by MATLAB.

4.3   Multiple Plots

IN SOME CASES IT MAY be desirable to present several plots in the same figure, provided that there is enough space to show the plots. We can easily achieve this with the following syntax:

```
> plot(x,y,'b-',x,sin(4*x),'k--')
```

The `plot` command can accept various plot specifications to show multiple plots.

Notice that the single `plot` command is taking two plot specifications, i.e. six arguments in total, three per plot. In the example above we are asking MATLAB and Octave to plot the $y = x$ function in a blue solid line and the $y = \sin(4x)$ in a black dashed line.

In cases like the one above it is highly recommended to include a legend that lets the reader know what is being plotted. This can be done by using:

```
> legend('cos(4x)','sin(4x)')
```

We can add a legend to the plot with the legend command.

The result of the command above will provide a list of line-styles, as they appeared in the plot command, followed by a brief description (the strings passed to the legend command).

The final output of the commands used above (including a title, labels and grid) can be seen in Figure 4.4. Please note that the figure created by MATLAB and Octave in the screen will have the graph for the $\cos(4x)$ in blue; the figure shown here is printed in black and white.

4.4 Holding Figures

EVERY TIME THAT THE WE create a graph with the plot command, MATLAB and Octave clear the contents of the current figure. In a lot of cases this is perfectly fine, however if we are interested in adding other plots to the same figure we can instruct the software to keep the current contents with the hold function. The command can be turned on or off as follows:

```
> plot(x,cos(4*x),'k-'), hold on
> plot(x,sin(4*x),'b--'), hold off
```

The current figure can be held with the hold command to continue adding plots to it.

The command hold on keeps the current picture whereas hold off releases it (but does not clear the window, which can be done with clf). Let us see what the above instruc-

To clear the window we can use the clf command.

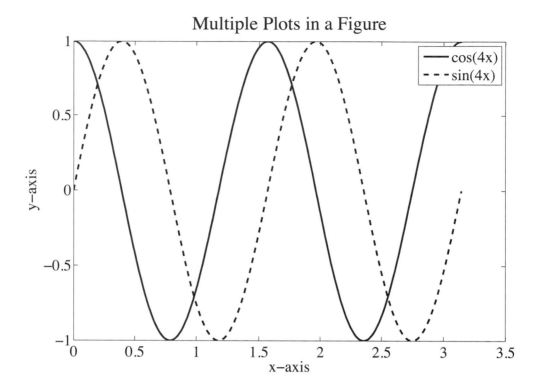

Figure 4.4: Multiple plots in the same figure can be placed using the plot command.

tions do: first we are asking the software to plot the vector that holds the values of the function

$$y_1 = \cos(4x) \tag{4.2}$$

against the values held by the vector x; the plot is done using a black solid line and we hold the figure for further plotting.

We then request the plot of the function

$$y_2 = \sin(x) \tag{4.3}$$

against the values of the vector x using a blue dashed line. Since we asked for the figure to be held, the second plot will be placed in the same frame as the previous one. Finally, we instruct MATLAB and Octave to free the plot with the command `hold off`. The result of the commands above can be seen in Figure 4.4, which shows the plots printed in black and white.

4.5 Subplots

IN THE PREVIOUS SECTION WE have seen how to add plots to the same figure window in a frame. However, in some cases we may prefer to split the graphics window to show plots in separate frames. We can do this by dividing the figure into an $m \times n$ array of smaller windows into which we may plot one or more graphs.

The sub-windows are counted 1 to mn row by row starting from the top left. Both `hold` and `grid` work on the current subplot and so these commands must be used as the relevant plots are being produced.

Subplots are enumerated by row from the top left.

The arrangement of subplots can be created with the `subplot(m,n,number)`, where m and n define the plot array and *number* is the label of the relevant subplot to use. For example, take a look at the following code:

An arrangement of subplots is achieved with the command `subplot(m,n,number)`.

```
> subplot(2,2,1)
> plot(x,cos(4*x))
> title('y=cos(4 x)')
> xlabel('x'), ylabel('cos(4 x)')
```

The command `subplot(2,2,1)` specifies that the window should be split into a 2×2 array and we select the first sub-window, where the function $y = \cos(4x)$ is being plotted using the default colour and line style.

We can continue filling in the rest of the other three subplots by issuing commands similar to the ones above while specifying the relevant number to refer to the correct frame where we want the graphs to be plotted. Look at the following commands:

```
> subplot(2,2,2)
> plot(x,sin(4*x))
> title('y=sin(4 x)')
> xlabel('x'), ylabel('sin(4 x)')
```

The subplot command requires us to tell it what subfigure in the array to use for plotting.

which will generate the second plot; for the third one we have:

```
> subplot(2,2,3)
> plot(x, x)
> title('y=x')
> xlabel('x'), ylabel('y')
```

and finally, for the fourth one:

```
> subplot(2,2,4)
> plot(x,x.^2)
> title('y=x^2')
> xlabel('x'),ylabel('y')
```

Notice that with each subplot command we specify what sub-window to use; for instance in the first case we have asked MATLAB and Octave to plot the function $y = \sin(4x)$ in the top right sub-window of the arrangement. Similar specifications have been made for the plots of $y = x$ and $y = x^2$ in the lower left- and lower right-hand corners of the main figure.

Each subplot can be identified with the index associated with it.

After plotting each of the subfigures we have issued commands to add appropriate labels to the axes. That is because those commands act only on the current subfigure. The result of the commands shown above can be seen in Figure 4.5.

Also, each subplot has its own specifications for title, labels, etc.

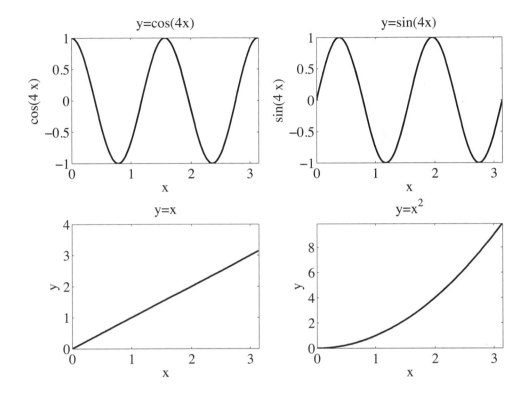

Figure 4.5: Subplots can be graphed with the command subplot. Each subplot can be given its own labels, grids, titles, etc.

4.6   Formatted Text

WHEN PRESENTING A PLOT, NOT only is it important to have a clear layout, supported by colours and line styles, but also with an appropriate legend, title and labels. The text shown in the plots can be further improved by using suitable format. For example, it is possible to increase or decrease the size of the text as well as add simple mathematical expressions (in LATEX form).

Let us take a look at an example by plotting in a panel the first 20 terms of the following sequence:

MATLAB and Octave accept simple LATEX commands to format text.

$$x_n = \frac{n^2 + 7}{n^2},$$ (4.4)

and then, in a second panel, plot the function

$$y_1 = \pi \cos^2(4x)$$ (4.5)

on the interval $-1 \leq x \leq 1$, finally adding some formatted text to the figure we are putting together. The tasks mentioned above can be achieved in MATLAB and Octave as follows.

In order to provide text that is readable, we can start by setting the default font size for the axis labels, legends and titles to a size of 16 points. In order to do that we can use the command set as follows:

```
> set(0,'Defaultaxesfontsize',16);
```

We can change the default font size.

The first argument issued above refers to the current figure, whereas the second one in this case tells the software what property we are changing, and finally we provide the new size for the font.

We can now define the series given by Equation (4.4). Since we are interested in the first 20 terms, we can define a vector containing the sequence of numbers from 1 to 20 and use that to calculate the series:

```
> n = 1:20;
> x = (n.^2+7)./n.^2;
```

This particular sequence converges to the value of 1, and we would like to show this in the plot we are going to create.

In the first panel we are therefore going to plot each of the 20 terms of the sequence marked by blue dots, and add a horizontal green dotted line to show the convergence of the sequence: The plot command is used to graph the sequence stored in the vector x versus the term number held in the vector n. In order to show the horizontal line, we create on the fly a couple of 2-element vectors whose values provide the information needed to plot the line. Finally, in the same plot command we use markersize to change the default value (6) of the marker size to the value of 10.

```
> subplot (2,1,1)
> plot(n,x,'.',...
     [0 max(n)],[1 1],'--','markersize',10);
```

The markersize property can be used to change the size of the markers in a plot.

Remember that the commands to add titles, labels and legends act only on the current figure. We add a title to the subplot, changing the size of the font to 20 using fontsize.

The fontsize property changed the size of the fonts used in a plot.

Similarly, we add labels and a legend. Notice that we can indicate to MATLAB and Octave the location of the legend with the aid of a second argument, where 1 stands for upper-right-hand corner, 2 is upper-left-hand corner, 3 is the lower-left-hand corner and 4 is the lower-right-hand corner.

The position of the legend can be specified with the help of the second argument passed on to the command legend.

MATLAB and Octave are able to understand some simple syntax used in LaTeX to generate mathematical formulae such as _, ^, \alpha, \pi, \sin, \cos, etc.

```
> title('x_n = (n^2+7)/n^2','fontsize',20);
> xlabel('n'), ylabel('x_n');
> legend('x_n','y = 1 ',1);
```

We can use LATEX syntax to format mathematical text in the plot.

The graph for the function $y_1 = \pi \cos^2(4x)$ is placed on a second subplot. The independent variable x is defined by a vector whose starting and ending values are given by the interval expected for this function. This vector is then used to evaluate the function y_1.

```
> subplot (2,1,2)
> x = -1:.02:1; y = pi.*cos(4*x).^2;
```

We can now plot the function, but this time let us change the thickness of the line with the help of linewidth.

```
> plot(x,y,'linewidth',2);
> legend('y_1 = \pi cos(4x)',1);
> xlabel('x'), ylabel('y_1');
```

We can change the width of the lines used in a plot with the linewidth property.

The result of the code explained above can be seen in Figure 4.6.

4.7   Changing Axes

ONCE A PLOT HAS BEEN created in the graphics window you may realise that a few changes are needed. Starting the plotting tasks from the start would be too time consuming, however there are ways to make changes to the plot even after it has been rendered. For example, you may wish to change the range of the x and y values shown on the graph.

Let us start by clearing the current figure window with the clf command and plot the function $y = x^2$:

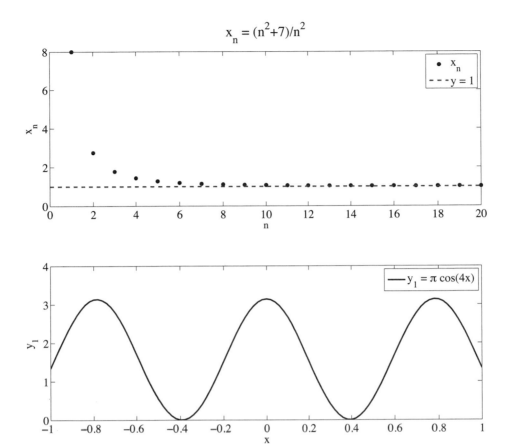

Figure 4.6: Multiple plots in the same figure can be placed using the plot command.

```
> clf;
> x=linspace(1,100);
> y=x.^2;
> plot(x,y)
```

We can modify the ranges of the plot as follows: Let us change the values of the x-axis to be between 10 and 20, and the y-axis to be between -2 and 500:

```
> axis([10 20 -2 500])
```

The axis command enables us to change the ranges of the values in the plot.

The axis command has four parameters: the first two are the minimum and maximum values to be used for the *x*-axis and the last two are the minimum and maximum values for the *y*-axis. In order to compare the changes let us enter the following commands in the software:

```
> clf;
> x=linspace(1,100);
> y=x.^2;
> subplot(1,2,1)
> plot(x,y)
> title('y=x^2 with default axes')
> xlabel('x'), ylabel('y')

> subplot(1,2,2)
> plot(x,y)
> axis([10 20 -2 500])
> title('y=x^2 with custom axes')
> xlabel('x'), ylabel('y')
```

The result of these commands is shown in Figure 4.7. We recommend taking a look at help axis as well as the information on the following:

- axis equal: the aspect ratio is set such that the data units are the same in each direction;

- axis square: makes the axes have equal lengths;

- axis normal: the aspect ratio is adjusted to fit the space in the figure;

- axis tight: sets the axes to the range of the data.

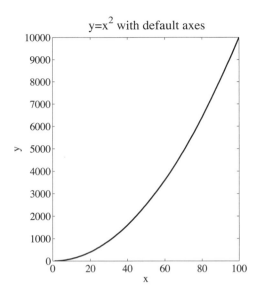

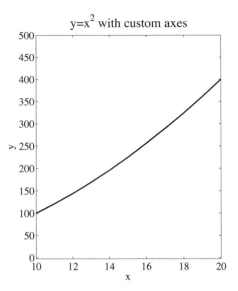

Figure 4.7: Graph of $y = x^2$ for values of x between 10 and 20. The y-axis is showing values between -2 and 500.

4.8 Plotting Surfaces

A SURFACE IS DEFINED BY a function $f(x,y)$; in other words, given the values of the pair (x,y), we can compute the value of the function by $z = f(x,y)$. The plot is thus a three-dimensional object.

In order to plot such a surface we have to decide on the ranges of both x and y. Let us for instance consider the intervals $1 \leq x \leq 14$ and $0 \leq y \leq 4$. This gives us a rectangle in the xy-plane. With these values we can generate a grid that contains each of the pairs given by the x and y values chosen. If we consider for instance a spacing of 0.25 between points, we can therefore define the vectors for x and y as follows:

```
> x = 1:0.25:14;
> y = 0:0.25:4;
```

The vectors above provide us with the length of each side of the square in the xy-plane, but we need to define the points that map that plane, i.e. a grid.

The grid can be defined in MATLAB and Octave with the command meshgrid, and the values obtained can be used to get the coordinates of the point $(X(i,j), Y(i,j))$ using the $i-$th point along from the left and the $j-$th point up from the bottom of the grid which correspond to the (i,j) entry in a matrix. The command meshgrid takes two vectors, x and y, that define each side of a rectangular grid resulting in a mesh of values that define the points inside it. Here is how:

The points used to define a plot surface can be calculated with the meshgrid command.

```
> [X,Y] = meshgrid(x,y)
X =
  1.000    1.250    1.500    1.750    2.000    . . .
  1.000    1.250    1.500    1.750    2.000    . . .
  1.000    1.250    1.500    1.750    2.000    . . .
  1.000    1.250    1.500    1.750    2.000    . . .
  1.000    1.250    1.500    1.750    2.000    . . .
  1.000    1.250    1.500    1.750    2.000    . . .

  . . .

Y =
      0        0        0        0        0    . . .
  0.250    0.250    0.250    0.250    0.250    . . .
  0.500    0.500    0.500    0.500    0.500    . . .
  0.750    0.750    0.750    0.750    0.750    . . .
  1.000    1.000    1.000    1.000    1.000    . . .
  1.250    1.150    1.250    1.250    1.250    . . .

  . . .
```

The meshgrid command generated the points inside a rectangular grid which in turn can be used to define a surface.

Please note that we have truncated the output shown above. In this particular case, the arrays X and Y are (17×53)-matrices and showing them here would not be very practical.

Finally, the values of the matrices X and Y can now be used to evaluate the function $f(x, y)$. Let us consider for instance the function

$$f(x, y) = \sin(x - 1) + \cos(3y - 1) \qquad (4.6)$$

for the ranges we had decided upon. The plot can be created in MATLAB and Octave with the surfl command:

```
> Z = sin(X-1)+cos(3*Y-1);
> surfl(X,Y,Z)
> colormap(gray)
> title('A surface')
> xlabel('x')
> ylabel('y')
> zlabel('z')
> grid on
```

The surfl command is used to create a surface plot.

The result can be seen in Figure 4.8. Notice that we used the command colormap to provide a colour scale to be used in the plot; some common colour maps in the software include:

The command colormap provides us with colour scales to be used in the surface plot.

- **Jet**: ranges from blue to red

- **HSV**: from red to red passing through yellow, green, cyan, blue, magenta

- **Hot**: from black to white passing through red, orange and yellow

- **Cool**: from cyan to magenta

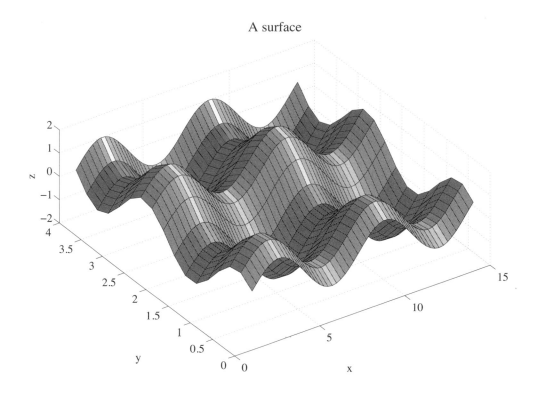

Figure 4.8: A surface plot obtained with the surfl command. Please note that this requires the generation of a grid with the command meshgrid.

- **Spring**: from magenta to yellow

- **Summer**: from green to yellow

- **Autumn**: from red to yellow

- **Winter**: from blue to green

- **Gray**: greyscale

- **Bone**: greyscale with higher values of blue

- **Copper**: from black to bright copper

- **Pink**: shades of pink

Instead of the command surfl we can use mesh, which instead of creating a surface filled with a smooth curve it defines the surface with a mesh only. You are encouraged to try the code above using the mesh command to see the result for yourself.

The mesh command generates a surface defined by a mesh instead of a smooth curve.

Sometimes it is useful to have a look at the projection of the surface on the xy-plane; you can release the command contour:

```
> contour(X,Y,Z)
> title('Countour of sin(x-1)+cos(3y-1)')
> xlabel('x'), ylabel('y');
```

The contour command lets us plot the projection of the surface on the xy-plane.

The result can be seen in Figure 4.9.

4.9 More Plots

WITH THE ELEMENTS WE HAVE discussed so far we can create a number of graphs, plots and surfaces that satisfy the needs of many applications and data visualisation tasks of our interest. In this section we will show a couple more plots that can be created with the help of MATLAB and Octave.

In the examples we have seen so far, the two-dimensional graphs shown have been plotted in linear Cartesian co-ordinates. Nonetheless, it is sometimes useful to be able to change the scale logarithmically or even plot in other coordinate systems altogether.

Not only can MATLAB and Octave produce plots in Cartesian coordinates but also in other coordinate systems and with different scales.

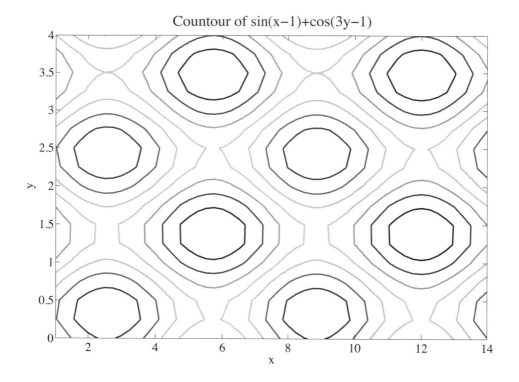

Figure 4.9: The projection of the surface shown in Figure 4.8. This plot was obtained using the `contour` command.

4.9.1   *Log Plots*

IT IS QUITE STRAIGHTFORWARD TO produce a plot that has a semi-logarithmic scale, and this can be done directly in the usual Cartesian coordinates and applying the familiar `plot` command. Let us for example visualise the function

$$f(x) = 4\exp(4x), \qquad (4.7)$$

for $0 < x < 10$ in a linear and a semi-logarithmic scale. To do that, let us create a vector for the values of x and define the function

```
> x = linspace(1,10);
> y = 4*exp(2*x);
```

In order to compare the two scales, we will use the `subplot` command to create an array of two graphs, one on top of the other:

```
> subplot(2,1,1);
> plot(x,y,'k','LineWidth',2);
> title('y=4 exp(2x) with linear axes')
> xlabel('x'), ylabel('y')

> subplot(2,1,2);
> semilogy(x,y,'k','LineWidth',2);
> title('y=4 exp(2x) with semilog axes')
> xlabel('x'), ylabel('y')
```

We can use the `semilogy` command to create plots in a semi-logarithmic scale.

The commands above are quite familiar to us, except for `semilogy`. Nonetheless, it is very easy to figure out what it does: it plots a graph in a semi-logarithmic scale. The output of the commands can be seen in Figure 4.10 and it is very easy to corroborate that in a semi-logarithmic graph, the function is represented as a straight line as shown in the bottom panel of the figure.

As we can see, in the top panel of Figure 4.10 it is quite difficult to see what is going on in the interval from 0 to 7. This is because the values are quite small. Nonetheless, by using the logarithmic scale, as shown in the bottom panel of Figure 4.10, it becomes much more obvious what is going on in that interval.

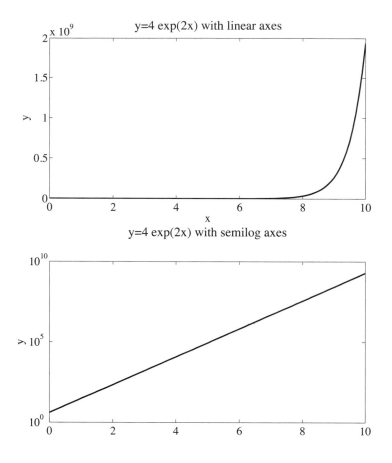

Figure 4.10: Comparison of the plot of the function $f(x) = 4\exp(2x)$ in linear Cartesian coordinates (top panel) and in a semi-logarithmic scale (bottom panel).

4.9.2   Plots in Other Coordinate Systems

MATLAB AND OCTAVE ARE ABLE to plot graphs in polar coordinates. This can be achieved with the polar command. For example, let us plot the cardioid function

$$r = 2\left(1 - \cos(t)\right) \qquad (4.8)$$

A cardioid is a plane curve generated by tracing a point on the perimeter of a circle rotating around a fixed circumference of the same radius. The name is given by the heart-like shape of the plot.

for values of $t \in [0, 2\pi]$. To do this, we will define a vector that contains the values of t and use them to create the plot as follows:

```
> t=0:0.01:2*pi;
> polar(t,2*(1-cos(t)))
```

We can plot the cardioid with the polar command.

The plot is effectively a curve described in the complex plane and the result of the commands above can be seen in Figure 4.11 where we are showing the output as produced by MATLAB; Octave will produce a similar, although simpler, graphic than the one shown here.

An alternative way to produce the plot is to generate the values of the cardioid function, but instead of creating the plot in polar coordinates we can find the appropriate transformation of these values into Cartesian coordinates. Let us take a look at how to achieve this. First we define the cardioid curve as follows:

```
> t=0:0.01:2*pi;
> r = 2*(1-cos(t));
```

We can now use the pol2cart command in order to find the Cartesian coordinates that correspond to the polar ones. The input for this command is the angular (t) and radial (r) coordinates and the output is the familiar x and y Cartesian coordinates.

The command port2cart finds the transformation from polar to Cartesian coordinates.

```
> [x, y] = pol2cart(t,r);
> plot(x,y);
```

Once the coordinate transformation has been carried out, we can use the command plot to visualise the graph and the result can be seen in Figure 4.12

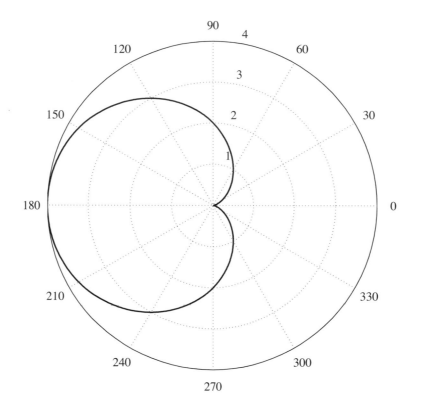

Figure 4.11: Polar plot for the function $r = 2(1 - \cos(t))$ as generated by MATLAB; Octave will produce a similar but simpler graphic. The heart-like shape of the plot inspired the name used for these functions: *cardioids*.

4.9.3   *Saving Plots*

CREATING FIGURES AND PLOTS IS a great tool while tackling a variety of problems and both MATLAB and Octave make it very easy to include the generation of graphs in any workflow. Nonetheless, in most cases, once we have some plots it is important to be able to save them or export them to formats that can be used in documents, reports, web pages and so on.

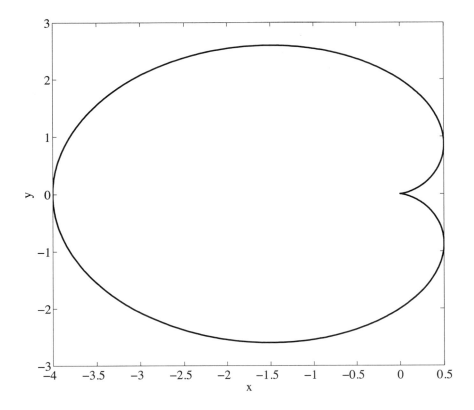

Figure 4.12: Plot for the function $r = 2\,(1 - \cos(t))$ in Cartesian coordinates. The coordinate transformation was handled by the pol2cart command.

We can export the figures we have created with the help of the print command which sends the contents of the current active figure window to the printer device. The command is able to take arguments to specify file formats or drivers, for example. For further information we recommend to take a look at the help for this command.

A general useful form of the command is as follows:

```
> print arg_1 arg_2 ... arg_n
```

The print command enables us to export our figures to the printer or to a file.

where arg_1 arg_2 ... arg_n are arguments passed to the print function. A typical form of the command is

```
> print -dformat filename
```

where -d*format* refers to the format of the file we want to save the figure to, and `filename` is the name of the file. Notice that the argument starts with the flag `-d` followed by the *format* desired; some common formats are listed in Table 4.2.

Format	Command
Encapsulated Postscript	`-deps`
JPEG	`-djpeg`
PNG	`-dpng`
PDF	`-dpdf`
Scalable vector graphics	`-dsvg`

Table 4.2: Some graphics formats supported by the command print.

For instance, if we require to save the current figure window to a PDF file with the name `myFigure.pdf` we can do so as follows:

```
> print -dpdf myFigure.pdf
```

and the file will be saved in the current path available to MATLAB and Octave.

4.10   Summary

ONE OF THE MOST DISTINGUISHING features of MATLAB and Octave is the capabilities they offer for visualising calculations and data immediately and within the same programming environment. We have seen how the software is able to use vectors and matrices, which we learned to manipulate on previous chapters, to plot functions in two and three dimensions.

The plot command lets us visualise the graphs of functions in two dimensions and is able to take multiple parameters to enable the visualisation of multiple graphs in one single figure. In order to render three-dimensional surfaces we can use the surfl and mesh commands in combination with the meshgrid command. Furthermore, several plots can be combined in a figure with the help of subplot.

Not only are we able to visualise data, but we can also add relevant and important information with the help of formatted text for titles, labels and annotations. This information can be made clearer with the capabilities that MATLAB and Octave have to process simple LaTeX commands. Finally, we have also seen how to change markers, font sizes, colours and line widths with the help of relevant parameters as well as saving our figures in different formats.

In the next chapter we will see how MATLAB and Octave combine the features we have discussed so far with scripting capabilities that allow us to explore data and models in a programmatical manner.

4.11   Exercises

1. Plot the function

$$f(x) = \frac{(6 + x^2)}{5 + 3x^2}, \qquad -4 \le x \le 5.$$

2. Plot the function

$$y = \exp\left(-x^2\right) \sin(\pi x^3), \qquad -2 \le x \le 2.$$

Provide a title and appropriate labels.

3. Plot the following functions in the same window for values between -2π and 2π. Make sure you include a legend and title, as well as labels for the axes. Each plot should have a different colour.

   (a) $y_1 = 3\cos(x)$

   (b) $y_2 = -2\cos(2x)$

   (c) $y_3 = -0.8\sin(x^2)$

4. Using the polar command, create a figure with the following subplots with $t \in [0, 2\pi]$:

   (a) $r_1 = e^{t/2}$

   (b) $r_2 = 2\sin^2(t) + 2\cos^2(t)$

   (c) $r_3 = 0.01t^2 - 9.02$

   (d) $r_4 = \cos(2t)$

   Compare the plots with those in Cartesian coordinates. Hint: Use pol2cart.

5. Create the surface and contour plots for the function

$$z = 2x^2 - 2xy + 6y^2,$$

   for $x \in [-25, 25]$ and $y \in [-20, 20]$

6. Plot the function shown below, including a formatted legend, labels and title:

$$f(x) = \sin\left(\frac{\pi}{2}x\right) + \sin\left(\frac{2}{5}\pi x\right), \qquad x \in [-3\pi, 3\pi].$$

7. Create the surface for the function

$$f(x, y) = (x^2 - y^2)\exp(-x^2 - y^2)$$

for x and y in the range $[-3, 3]$.

8. Consider the set of parametric equations:

$$X = 1.5\cos 4t, \qquad\qquad Y = 2\sin 5t.$$

Create the following plots:

(a) X versus t

(b) Y versus t

(c) Y versus X

(d) Find out about the built-in function comet and use it to animate the Y versus X plot.

9. Plot the following functions in rectangular, semilog and log-log scales, with $x \in [0, 3]$:

(a) x

(b) x^2

(c) e^2

(d) e^{x^2}

Hint: find out information about loglog.

10. Plot the function $f(x) = x + \cos(x^5)$ with x=−5:0.1:5. What is wrong with this plot? How can you correct it? Produce subplots for the two graphs in a single figure and print it as a jpeg file.

5
Programming MATLAB® and Octave

UP UNTIL NOW WE HAVE been using MATLAB and Octave in a way where an input is immediately followed by its output. Although this is generally fine and useful to a certain extent, in practice we are required to repeat and change commands. Similarly, once a sequence of commands is able to perform a determined task we are quite likely to use it again and thus we need to store it. Not only does this simplify our workflow but also helps in the reproducibility of results. All these tasks can easily be achieved with the use of scripts and programming directives. We shall talk about them in this chapter.

Scripts and programming commands can help simplify our workload.

5.1 *Script Files*

SCRIPT FILES ARE SIMPLE TEXT files that contain and store MATLAB and Octave commands. These commands are indeed those that we have described in the previous chapters, with the added advantage that we can define logic for their execution; in other words we can use these files to program MATLAB and Octave.

As mentioned above, script files are composed of unformatted text and an easy way to distinguish them in our file system is thanks to their extension: *.m*. For example, we can in general tell that a file called `MyScript.m` is quite possibly a script that can be executed by MATLAB and Octave. Thanks to the extension used, these files are commonly known as m-files. The commands in any of these files can be executed by typing the name of the file itself in the command line; the extension is not needed for the script to be executed.

Scripts for MATLAB and Octave have the extension *.m*.

5.1.1   *Text Editors*

RUNNING AN M-FILE RESULTS IN the commands contained in the script to be executed and, in case it is required in the programme, their output displayed on the screen. Scripts or m-files can be created with your favourite editor as long as they are saved as simple text. This is important as MATLAB and Octave will read the script line-by-line and execute the commands as they are encountered. If you use an editor that adds formatting commands (such as Microsoft Word), then the software will fail to run your script. In the case of MATLAB, you can use the script editor that is included with the software: simply click on the "New Document" icon at the top left of the main MATLAB window. Once you are in the editor, you just need to type the commands needed and then save the file (remember that it should have a .m extension). For Octave in a Macintosh environment a number of users recommend editors such as *Textwrangler*, *Aquamacs* or *Sublime*; whereas in windows *Notepad++* is very useful. Finally, for Linux/Unix editors such as *Emacs*, *Nano* or *gEdit* can be used to edit scripts. All these editors can also be used in conjunction with MATLAB, too.

Ensure that your scripts are saved as simple text.

Text Editors:
- **Textwrangler**: `http://www.barebones.com/products/textwrangler/`
- **Aquamacs**: `http://aquamacs.org/`
- **Sublime**: `http://www.sublimetext.com`
- **Notepad++**: `http://notepad-plus-plus.org/`
- **Emacs**: `http://www.gnu.org/software/emacs/`
- **Nano**: `http://www.nano-editor.org`
- **gEdit**: `http://gedit.en.softonic.com`

5.1.2 *Adding Comments*

IT IS A GOOD PRACTICE to write comments that explain
what it is that you are trying to achieve with the flow of
your programme. This will help you, and anyone else using
your code, follow each bit of the programme and make
sense of the code. In order to create comments you use the
% symbol. Any text that follows this symbol is treated as a
comment and is not executed by MATLAB and Octave. This
is very useful for debugging and trying different commands
in your scripts: since a commented line is not executed,
the % symbol can be used with good effect for telling the
software not to run one or more lines of code, without
having to delete them. Should you require them again later
on, all you have to do is delete the commenting symbol.

Add comments to your code: any
lines that start with the % symbol
are not executed.

5.2 *Flow of a Programme*

MATLAB AND OCTAVE USE PROCEDURAL programming
to execute the scripts. This means that whenever we run
a programme we invoke procedures, routines, methods or
functions that contain a series of computational steps to
be carried out one after the other. You can think of a script
as a list of instructions that you are asking the software
to execute, and this is done on a line-by-line basis as we
mentioned earlier. In other words, the execution of the
m-file is done from the top to the bottom of the script
sequentially.

A script is executed sequentially,
line by line.

Although this is an easy procedure to understand, there
are times when we require the programme to follow a
particular list of instructions depending on the outcome
of the procedures executed earlier on in the script, or to

repeat a certain procedure either a certain number of times or until a condition is met. These decisions can be taken with the help of Boolean operations that MATLAB and Octave understand. In that way we can decide if something is true or false, represented as 1 or 0, respectively, and take appropriate action.

Boolean operations result in values TRUE or FALSE.

5.2.1   Relational Operators

IF AT SOME POINT IN a calculation a variable x has been assigned a value, it is possible to make certain logical tests on it: for instance we may check if the value held by x is equal to a particular number, for example, 42:

```
> x=42;
> x==42


ans =

     1
```

A logical test for equality is run with two equal signs: ==.

Relational Operator	Meaning
==	equal to
~=	not equal to
>	greater than
<	less than
>=	greater than or equal to
<=	less than or equal to

Table 5.1: Some relational operators supported by Matlab and Octave.

Please note that we are using two equal signs as this is a logical test, not a value assignment, i.e. we are not assigning the value of 42 to the variable x as done on the first line of the example above. Other tests that can be run include

comparisons to see if the value of the variable x is greater
than or less than a given number. Table 5.1 shows some of
the logical operators supported by MATLAB and Octave.
Let us have a look at an example.

Let us consider assigning the value of 2π to the variable x.
we do this with a single equal sign:

```
> x = 2*pi

x = 6.2832
```

We can then ask the software to check if x is not equal to 6.
From Table 5.1 we know that the operation "not equal to"
can be carried out with the operator ~=:

```
> x~=6

ans =

        1
```

If the result of a comparison
is TRUE the value returned by
MATLAB and Octave is 1.

The result of the comparison is 1 and thus we know that the
comparison is TRUE, in other words the value held by the
variable x is indeed not equal to 6. Let us see what happens
when we ask if the value held by the variable is not equal to
2π:

```
> x~=2*pi

ans =

        0
```

If the result of a comparison
is FALSE the value returned by
MATLAB and Octave is 0.

In this case the result is FALSE and thus the software returns the value of 0. The comparisons do not necessarily
have to be made with a variable; they can also be made
directly with numbers:

```
> 42>=24

ans =

      1
```

as well as with the results of calculations:

```
> pi^2<100

ans =

      0
```

5.2.2   Relational Operators with Vectors and Matrices

As we know, the default type of object handled by
MATLAB and Octave is a matrix, and as such it is natural
to ask how these logical operators are executed on them.
When the object is a vector or a matrix, these tests are done
on an element-by-element basis.

Relational operations can also be
used with vectors and matrices.
The comparisons are made
element by element.

For instance, if we consider the matrix:

$$x = \begin{pmatrix} 0 & 1 & 2 & 3 & 4 \\ 8 & 2 & 0 & 0 & 0 \end{pmatrix}, \qquad (5.1)$$

and we are interested to know what elements are equal
to zero, we can obtain a Boolean matrix indicating those
elements where the comparison is TRUE with a value of 1
and those where the comparison is FALSE with a value of 0.
Let us enter the matrix shown in Equation (5.1) and make a
comparison with 0:

The result of a relational operator
on a matrix is a Boolean matrix
where each element is the result of
the comparison on an element-by-
element basis.

```
> x=[0:4; 8 2 0 0 5];
> x==0

ans =
     1     0     0     0     0
     0     0     1     1     0
```

We have asked the software to evaluate if the values of the matrix x are equal to 0 or not; that is why the answer matrix contains the number 1 (TRUE) in the places where the original matrix has entries with the value of 0.

We can test for other values, for example we can obtain a matrix that tells us what entries are strictly greater than 1.

```
> x>1

ans =
     0     0     1     1     1
     1     1     0     0     1
```

We can carry out comparisons for different values other than 0.

Similarly, we can ask for values that are greater or equal to 1 with the use of the >= operator as follows:

```
> x>=1

ans =
     0     1     1     1     1
     1     1     0     0     1
```

The comparison can be done with any of the relational operators shown in Table 5.1.

As shown above, the usage of relational operators is the same regardless of the type of object we are comparing. This means that we can carry out the comparison of one matrix to another, or a vector to another. In order for the comparison to make sense, the only thing we have to take into account is the fact that the matrices or vectors to be compared have the same dimensions.

For example, we can compare the following two matrices:

$$x = \begin{pmatrix} 17 & 24 & 1 & 8 & 15 \\ 23 & 5 & 7 & 14 & 16 \\ 4 & 6 & 13 & 20 & 22 \\ 10 & 12 & 19 & 21 & 3 \\ 11 & 18 & 25 & 2 & 9 \end{pmatrix}, \qquad (5.2)$$

$$y = \begin{pmatrix} 1 & 2 & 3 & 4 & 5 \\ 6 & 7 & 8 & 9 & 10 \\ 11 & 12 & 13 & 14 & 15 \\ 16 & 17 & 18 & 19 & 20 \\ 21 & 22 & 23 & 24 & 25 \end{pmatrix}, \qquad (5.3)$$

Comparisons using the relational operators from Table 5.1 also work with matrices and vectors.

and we are interested to know what elements of x are greater than those of y.

We can do this as follows: the first matrix, x, is in fact a 5×5 magic matrix:

```
> x=magic(5)

x =

    17    24     1     8    15
    23     5     7    14    16
     4     6    13    20    22
    10    12    19    21     3
    11    18    25     2     9
```

whereas the second one can be constructed with sequences.

Let us do this by constructing a vector

$$v = \begin{pmatrix} 1 & 2 & 3 & 4 & 5 \end{pmatrix}, \qquad (5.4)$$

and use this to construct the matrix by adding multiples of 5 each time:

```
> v = [1:5]

v =

     1    2    3    4    5

> y=[v; v+5; v+10; v+15; v+20]

y =

     1    2    3    4    5
     6    7    8    9   10
    11   12   13   14   15
    16   17   18   19   20
    21   22   23   24   25
```

We can now carry out the comparison as follows:

```
> x>y

ans =
     1    1    0    1    1
     1    0    0    1    1
     0    0    0    1    1
     0    0    1    1    0
     0    0    1    0    0
```

The result of the comparison of two $m \times n$ matrices results in a Boolean matrix of the same dimensions.

Notice that the result of the comparison is indeed a 5×5 matrix with entries equal to 1 where the comparison is true, and 0 where the comparison is false.

5.2.3 *Logical Operators*

LOGICAL OPERATORS ALLOW US TO create more complex tests than the simple comparison carried out with the relational operators discussed above on their own. For instance, if we need to check whether a variable is both greater than 1 and less than 42, we can use the logical operator & (meaning AND) to ensure that the value held by the variable is between 1 and 42 (but not equal to either).

```
> x = 20;
> x > 1 & x < 42


ans =
        1
```

The logical operators in MATLAB and Octave follow the usual truth tables used in logic.

Some logical operators supported by MATLAB and Octave are listed in Table 5.2.

Operator	Meaning
&	AND
|	OR
~	NOT
&&	Short-circuit AND
||	Short-circuit OR

Table 5.2: Some logical operators supported by MATLAB and Octave.

Notice that we have listed operators described as "short-circuit". The difference between the normal logical AND and OR and their short-circuit versions is that the latter employs a behaviour that halts the evaluation of the comparison if the first expression enables the truth value of the statement to be determined. For example, consider the following line of code:

```
> (4>10) && expression1
```

Since 4 is smaller than 10, the evaluation of the comparison (4>10) will return the value false and thus, the overall truth value of the full statement is false, irrespective of the truth value of expression1.

THE COMPARISONS USING RELATIONAL OPERATORS together with the logical ones are still carried out on an element-by-element basis. Let us look at some more examples. Consider the following matrix:

$$x = \begin{pmatrix} 0 & 1 & 2 & 3 & 4 \\ 8 & 2 & 0 & 0 & 5 \end{pmatrix}. \tag{5.5}$$

We can enter this matrix in MATLAB and Octave as follows:

```
> x = [0:4; 8 2 0 0 5]

x =

   0   1   2   3   4
   8   2   0   0   5
```

We can then ask for the entries in this matrix that are greater than 3 and less than 5:

```
> (x > 3) & (x < 5)

ans =

     0     0     0     0     1
     0     0     0     0     0
```

The logical operators carry out comparisons on an element-by-element basis.

The brackets in the expression used in the example above are not necessary, but they help with reading the combination of commands entered in the software. We can also ask for the entries in the matrix shown in Equation (5.5) that are greater than 3 or equal to 8:

```
> (x > 3) | (x==8)

ans =

      0     0     0     0     1
      1     0     0     0     1
```

Remember that the result of a logical comparison is a Boolean matrix, where 1 represents the value for TRUE and 0 the value for FALSE.

Remember that the logical test for equality is expressed with the double equal sign as shown in the example above.

5.2.4   Selecting Elements with Logical Operators

SINCE THE LOGICAL OPERATORS WE have discussed are applied on each element of a vector or matrix, it is possible to use them in order to select components that do not meet certain criteria. This is effectively achieved by multiplying, element-by-element, the original array by a logical one of the same dimensions.

Let us for example generate a vector with $n = 5$ elements drawn from a Gaussian distribution with mean $\mu = 1$ and standard deviation $\sigma = 2$.

The randn(m,n) command generates an $m \times n$ matrix whose elements are uniformly distributed pseudorandom numbers.

The randn command generates arrays of uniformly distributed pseudorandom numbers.

```
> n=5;
> mu=1, sigma=2;
> x = mu + sigma*randn(1,n)


x =
   -0.58461   -2.32660    0.93833    3.19227    3.52279
```

Should you try this code in your computer, you will obtain different values from the ones shown above.

We can now obtain an array that tells us where the values of the matrix x are smaller than $\mu - \sigma$:

```
> x<mu-sigma


ans =
      0   1   0   0   0
```

Logical operators are useful to identify elements that fulfil a condition.

In the example above, only the second element is smaller than the value we were checking for and now we are in a position to select it. We can do this by multiplying the original matrix by the logical one on an element-by-element basis:

```
> x1=x.*(x<mu-sigma)


x1 =
   -0.00000   -2.32660    0.00000    0.00000    0.00000
```

If instead, we wanted to remove this value, we can carry out a similar operation, but this time the logical operator would be for a greater-than relationship:

Together with matrix multiplication, logical operators can help us select elements we are interested in.

```
> x2=x.*(x>mu-sigma)

x2 =
   -0.58461   -0.00000    0.93833    3.19227    3.52279
```

In the example outlined above, we have used a small number of elements and thus it is very easy to determine which of them are the ones that fulfil the condition tested for. However, in larger vectors or matrices this may not be so straightforward. In those cases, it is possible to use the find command in conjunction with logical operators to identify those elements. For example, let us consider a similar problem as the one above, except that this time we construct a 5×5 matrix with normally distributed elements with the same parameters as in the previous example:

With larger arrays, the use of the find command can be a good way to identify elements that meet a condition.

```
> n=5;
> mu=1, sigma=2;
```

In this case we will be using the parameters passed to the randn command to construct a square matrix with the desired number of elements:

```
> x = mu + sigma*randn(n,n)

x =
    2.0753   -1.6154   -1.6998    0.5901    2.3430
    4.6678    0.1328    7.0698    0.7517   -1.4150
   -3.5177    1.6852    2.4508    3.9794    2.4345
    2.7243    8.1568    0.8739    3.8181    4.2605
    1.6375    6.5389    2.4295    3.8344    1.9778
```

The find command returns the linear indices of the non-zero elements of a matrix and thus we can find for example the elements that are smaller than $\mu - \sigma$ as follows:

The command find returns the linear indices of the non-zero elements of an array.

```
> mu-sigma

ans =

     1

> cond=find(x<mu-sigma)


cond =

     3

     6

    11

    22
```

The indices returned by the command `find` are ordered column-by-column.

Notice that the linear indices are listed on a column-by-column basis; in the example above the first element that meets the condition tested for is in row number 3 of the first column. The second one is the sixth element of the matrix, and since this is a 5×5 array, then element number six is the first element of the second column, and so on.

Remember that the example used above used the `randn` command and the numbers shown here will differ from those you may find when running this command in your computer.

ANOTHER USEFUL APPLICATION OF THE logical operators in MATLAB and Octave is in the plotting of functions. For example, let us consider plotting the following function:

Logical operators can be put to good use in generating plots.

$$y = \tan(x), \text{ for } -\pi \le x \le \pi. \qquad (5.6)$$

Using the discussions from Chapter 4, we can plot this function with the following commands:

```
> x = [-pi:pi/100:pi];
> y = tan(x);
> plot(x,y)
```

We have simply generated an array with suitable values for x and used them to calculate the tangent. The result of the plot can be seen in Figure 5.1. As we can see, the function obtained does not look anything like the tangent function. This is because the tangent has singularities at $x = \pi/2$ and $x = -\pi/2$ in the interval chosen. Quite understandably, MATLAB and Octave do their best to generate a large value at these points to mimic infinity at these points, but instead of showing the usual shape of this function we end up with a bad figure.

One way to avoid this situation is by eliminating the very large values in the data. This can easily be done with a logical operator as follows:

```
> y = y.*(abs(y)<1e10);
> plot(x,y)
```

What this operation does is to calculate the absolute value $|y|$ with the command abs and those values that meet the condition that $|y| < 10^{10}$ get multiplied by 1, whereas in the cases where $|y| > 10^{10}$ the values are multiplied by 0. The result of this extra operation can be seen in Figure 5.2, where it is now possible to see the distinctive features of the tangent function, which were completely masked by the large values obtained when not using the logical operators, as shown in Figure 5.1.

The abs command calculates the absolute value of an object.

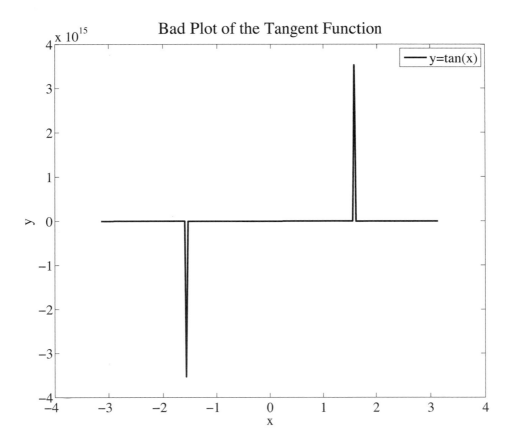

Figure 5.1: Plotting functions with singularities leads to figures that do not represent the true characteristics of the function. In this case we are showing what happens when trying a naive approach when plotting the tangent function $y = \tan(x)$.

5.3   *Loops in MATLAB and Octave*

WE MENTIONED BEFORE THAT IT is sometimes desirable to repeat a certain number of commands depending on values obtained in the course of the execution of a programme. Furthermore, in the previous section we have seen how a comparison between values can be carried out. In this section we will address how the repetition of steps can be achieved.

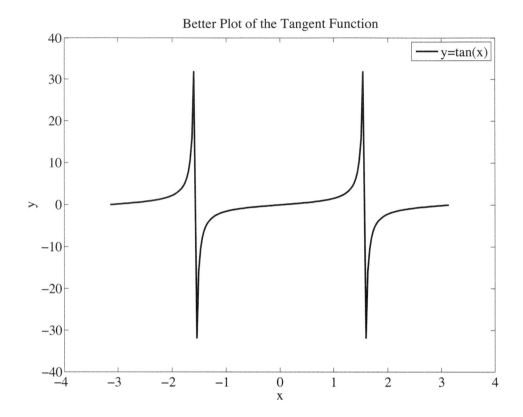

Figure 5.2: The use of logical operators can help us improve those plots where singularities may arise. Here we can indeed distinguish the important features of the tangent function $y = \tan(x)$.

In programming, a loop is a sequence of commands that is continually repeated until a certain condition is met. Generally, the programme carries out a process and, with the help of logical and relational operators, we can check if the condition imposed has been fulfilled. For instance, we can check if a certain value has been obtained or a maximum number of repetitions has been executed. If the condition is not fulfilled, the sequence is repeated once again and the condition checked again. Something that should be avoided when writing code is the occurrence of a so-called *infinite loop*, i.e. a sequence of steps that lacks a

A loop is a sequence of steps that is repeated until a condition is met.

Avoid the creation of infinite loops.

functioning exit routine. As we shall see below, there are different kinds of loops.

5.3.1   *For Loop*

IN SOME PROCEDURES IT IS necessary for the software to execute a particular instruction a certain predetermined number of times. This can easily be achieved with a for loop. The syntax of this type of loop in MATLAB and Octave is as follows:

```
for index = initialval:finalval
      procedure to be executed for each
      value in the loop
end
```

For loop: the steps are repeated for values of index starting at initialval and finishing at finalval. In Octave it is possible to finish the block with **endfor**.

In the syntax above index is the control variable against which a logical operation is carried out. The index will first take the value given by initialval and the loop will stop when the value finalval is reached; for every value taken by index the procedure inside the loop will be executed. Let us have a look at a simple example. Consider the vector

$$v = (1, 4, 7, 10), \tag{5.7}$$

which can be produced with a sequence that starts with number 1 and finishes with number 10 on steps of value 3:

```
> v = [1:3:10]

v =

      1   4   7   10
```

Imagine now that we require to change these values with a linear sequence from 1 to 4. Please note that we can simply reassign the variable v with the sequence outlined above; nonetheless this is a simple example to illustrate the use of a `for` loop. We can do this as follows:

```
> for j=1:4, v(j)=j; end
> v


v =

      1  2  3  4
```

The `for` loop syntax requires the use of end to terminate the loop.

In this example we have started with the vector given by Equation (5.7) and used a `for` loop to reassign its elements to the values of the index j going from 1 to 4. This is a simple demonstration of how this loop works, but it is not very practical given the procedural nature of the software, as looping through a vector in this way is very slow. Instead we could have used a syntax similar to the first statement in the example above and which is more natural in both MATLAB and Octave.

In the example above the `for` loop moved in steps of 1 from the initial value to the end value; we can also require that the sequence moves with different step values. In this case, the syntax of the loop makes use of the colon notation we are familiar with:

```
for index = initialval:step:finalval
     procedure to be executed for each
     value in the loop
end
```

We can use the colon notation to change the step in the `for` loop.

The loop will start at the value given by `initialval` and every iteration will be executed until the value `finalval` is reached; this will be done by changing the `index` on every iteration by the value given by `step`.

For example, let us consider the case where we want to generate the following vector with a for loop:

$$v = (1, 3, 5, 7).\qquad(5.8)$$

We can do this as follows: let us start a counter variable k that controls the elements of the array and as such we initialise it with the value of 1. We can then set up a for loop starting at 1 and finishing at 8, changing the index j with a step equal to 2. Before each iteration is completed we add one to the control variable k so that the next value is stored in the correct place:

```
> k=1;
> for j=1:2:8 ...
    v(k)=j;...
    k=k+1;...
  end;
```

The use of a control variable that is changed within the loop gives us the flexibility we may require in constructing loops.

We can easily check that the values stored in the vector v are as expected:

```
> v

v =

        1    3    5    7
```

The for loop used to construct this vector is not very practical. Vectorisation should instead be used.

Before we move on, let us mention a couple of points. It is usually the case that programmers use the variables i, j or k to denote the controls in a loop. This is indeed true in the case of MATLAB and Octave, too. Nonetheless, we urge some caution as the i and j are used in the software to denote the complex number i.

Remember that i and j are also used to represent the complex number i.

Also, in the example above, you would have noticed that the for loop is made up of four lines of code, each of which finished in an ellipsis, i.e. ..., which is required by MATLAB to signify the continuation of a command. However, Octave does not have this requirement and thus the code above can be written as:

```
> k=1;
> for j=1:2:8
    v(k)=j;
    k=k+1;
  end
```

Octave

We shall continue using the ellipsis in the rest of the book to avoid confusion.

5.3.2   While Loop

THERE ARE SOME OCCASIONS WHEN we want to repeat a section of code until some logical condition is satisfied, but unlike the case of the for loop explained in the previous section, we may not be able to tell in advance how many times we have to go around the loop. This can be done with a while loop. The syntax of this loop is

```
while expression
    procedure to be executed
    while the condition given
    by the expression is satisfied
end
```

The procedure in a while loop is executed while the expression tested is satisfied. In Octave it is possible to finish the block with **endwhile**.

In this case expression is a logical test that can be evaluated to be either true or false. The procedure will therefore be executed while the expression is true and as soon as it is false, the loop will stop. For example, starting with the value of 0, we would like to print a sequence of numbers less than 3. This can be done with the following code:

```
> n=0;
> while n<3 ...
      n = n+1 ...
   end

n =

      1
n =

      2
n =

      3
```

The while loop tests the condition before each iteration is carried out.

Notice that we need to change the value of the control variable n inside the while loop; this helps us avoid the creation of an infinite loop as eventually the condition is no longer satisfied. Note also that the logical test is carried out at the beginning of the loop. This is important as the loop may not even execute once if the condition is not met. For example, consider the following code:

```
> n=10;
> while n<3 ...
      n = n+1 ...
   end
```

If the condition is not met at the beginning of the while iteration, the code inside the loop is not executed.

Since the initial value of the control variable n is 10, the first time the condition is tested results in a FALSE value and the loop is not executed.

IT CAN BE VERY HELPFUL to use a while loop in programming flows that require that a certain condition is met before the next iteration takes place. Let us look at an example. Consider solving the following equation numerically:

$$x = \cos^2(x). \qquad (5.9)$$

We can start with a guess of say $\pi/2$ and then compute a sequence of values for $x_n = \cos^2(x_{n-1})$, where n is the number of iterations, and continue until the difference between consecutive values x_n and x_{n-1} is smaller than a tolerance given. We can start by defining some useful parameters in the solution of Equation (5.9):

```
> x_guess = pi/2;
> n=1;
> difference=1;
```

where x_guess is our initial guess, n assigns the value for the first iteration and we define an initial difference for comparison purposes. We can now set up a while loop:

```
> while (difference > 0.001) && (n< 200) ...
    n = n+1; ...
    x_new=cos(x_guess)^2; ...
    difference = abs(x_new - x_guess); ...
    x_guess = x_new; ...
  end

> n
n =
    146
> x_guess
x_guess =
    0.6412
```

The while loop can be very useful to test that a certain condition is met before the next iteration takes place.

Notice that at the beginning of the loop shown above we have written a line that checks first if the tolerance of 0.001 has been met. Also, since we do not want the programme to continue indefinitely (in case the tolerance is not met), we have put a maximum number of iterations (200 in this case). The difference between the new value and the guess is tested. Note also that the new value replaces the guessed value every time we go around the loop. The actual answer for this problem is approximately 0.6417, so our loop is giving us a good estimation for the answer of Equation (5.9).

5.4   Conditionals: If... Then... Else...

As we have seen, making a decision within the flow of a programme is very helpful. Sometimes that decision is not just related to the number of times a particular procedure is repeated, as in the examples in the previous section, but it might depend on whether a logical operation returns a true or a false statement.

If the statement is TRUE, we would like our programme to carry out a specific procedure, but if it is FALSE a different path should be followed. The syntax for this type of decision is as follows:

```
if logical test
      procedure executed
      when the logical test is TRUE
else
      procedure executed
      when the logical test is FALSE
end
```

The if statement enables us to take different paths in the programme depending on the result of a logical test. In Octave it is possible to finish the block with **endif**.

The logical test referred to above is an expression that uses the relational and logical operators defined in Sections 5.2.1 and 5.2.3, respectively.

Let us have a look at a simple example. Given the expressions $\cos(\pi/3)$ and $\sin(\pi/3)$ we want to write some code that tests which one is larger and display a message accordingly. We can use the if... then... else statement as follows:

```
> a = cos(pi/3); b = sin(pi/3);
> if a > b ...
    disp('cos(pi/3) is bigger than sin(pi/3)')...
  else ...
    disp('sin(pi/3) is bigger than cos(pi/3)')...
  end

sin(pi/3) is bigger than cos(pi/3)
```

We are using the disp command to print a message to the terminal.

In the example above we have used the command disp to display a message to the terminal. In this case, since

$$\sin\left(\frac{\pi}{3}\right) = \frac{\sqrt{3}}{2} \simeq 0.8660 \qquad (5.10)$$

is a bigger value than

$$\cos\left(\frac{\pi}{3}\right) = \frac{1}{2} = 0.5, \qquad (5.11)$$

then the logical test is not TRUE and therefore only the statement after else is displayed.

In other cases there may be more than one path that the programme should follow, depending on whether a set of conditions is either TRUE or FALSE. In those cases a nested if... then... else... can be used; however, there is a shortcut offered by MATLAB and Octave:

```
if logical test1
    procedure executed
    when the logical test1 is TRUE
elseif logical test2
    procedure executed
    when the logical test2 is TRUE
    ...
else
    procedure executed
    when all the other logical tests
    are FALSE
end
```

MATLAB and Octave support the use of nested if statements using the command elseif followed by an additional logical test.

In this case we can include as many logical tests as we want or need. However, it is important to note that as soon as one of the logical tests returns a TRUE value, the other logical tests will be ignored. We should therefore be very careful in the order in which the tests are performed. Finally, if none of the logical tests returns a TRUE value, then the procedure following else will be executed. For instance, if we want to see which of $\cos(\pi/4)$ or $\sin(\pi/4)$ is bigger we can write the following:

As soon as one of the logical tests is TRUE, the rest of the logical tests are not checked.

```
> a = cos(pi/4); b = sin(pi/4);
> if a-b > 0.0001 ...
  disp('cos(pi/4) is bigger than sin(pi/4)')...
  elseif a-b < -0.0001 ...
  disp('sin(pi/4) is bigger than cos(pi/4)')...
  else ...
  disp('sin(pi/4) is equal to cos(pi/4)') ...
  end

 sin(pi/4) is equal to cos(pi/4)
```

If none of the logical tests in a nested if return TRUE, the procedure after the command else is then executed.

We have written the code in this way, i.e. providing a tolerance for the comparison, in order to avoid the fact that there are rounding errors when computing the values required. Since neither the first not the second logical tests were TRUE, the code displays the message after the else statement.

5.5 Procedures and Functions with m-Files

AS MENTIONED EARLIER ON, UP until now we have been using MATLAB and Octave effectively as sophisticated line-by-line calculators. However, they can be used in a much more powerful way by writing scripts. This allows us to create more complicated programmes and automate a number of tasks as well as repeat calculations by running the script, instead of typing commands again and again. We know that these scripts are referred to as *m-files*. When an m-file is run, the commands written in the script are interpreted one by one as explained in Section 5.1.

An *m-file* is a simple text file that enlists a series of commands to be executed. If the script accepts inputs and returns outputs then we call it a function.

When these m-files are used for executing specific tasks given an input and the result is an expected output, then we refer to *function m-files*.

5.5.1 Putting It All Together: m-Files

IF WE WANT TO EXECUTE repeatedly a given set of commands, we can create an m-file that lists these commands and stores them for later use. For example, let us consider a situation where we want to be able to plot a sine or cosine function with a given frequency between $-\pi$ and π.

We can indeed use the terminal as we have done so far, but as soon as we want to change the frequency of the function, for instance, then we have to re-type all the different commands. However, we can create a script that can be saved in our computer and later retrieved. All we would have to do then is to test whether we want to plot the sine or the cosine, modify the frequency of the trigonometric function to be plotted, run the script and we are done.

One important thing to remember is that unlike in the examples we have seen in the previous sections, the scripts do not need to end each line in the code with an ellipsis, unless we truly are splitting a single long command.

m-files do not need the use of an ellipsis (...) at the end of each line.

Let us tackle the creation of the m-file that can be used for the task of plotting either the sine or cosine functions as mentioned above. First we will create a text file that will hold the commands and instructions so that they are readily available at a later stage. It is important to mention that in order for MATLAB and Octave to execute the script, you must be in the folder or directory where the file is saved. In this case we will name our script `my_plot_script.m`. Remember that you can use your favourite text editor as explained in Section 5.1.1.

It usually helps to have a brief plan of the tasks we need to accomplish to tackle our problem. For instance, in this case we can take into account the following list:

1. Make sure that the workspace is clear.

2. Define a variable to hold the frequency of the trigonometric function.

3. Define the range of values to plot the trigonometric function.

4. Decide if the sine or the cosine will be plotted.

5. Create the plot and format it.

The following script accomplishes these tasks:

```
% Clear the memory
clear;
% Define the frequency of the trigonometric
% function
frequency=1;
% This variable will allow us to make
% a decision about what function
% will be plotted
whichplot=1;
% whichplot = 2;

% Define the range of the plot
x=linspace(-pi,pi,100);
% Decide which function will be plotted
if whichplot==1
    % Sine if whichplot is one
    plot(x,sin(frequency*x),'k-.')
    title('Sin')
    xlabel('x'), ylabel('y')
else
    % otherwise the cosine
    plot(x,cos(frequency*x),'k--')
    title('Cos')
    xlabel('x'), ylabel('y')
end
```

The % symbol is a comment; this line is not executed.

We define the variable frequency that can later be changed to modify the behaviour of the programme.

We define a variable whichplot which enables us to change the path the programme can take and thus plotting either the sin or the cos functions.

The very first line of the script starts with a percentage (%) sign. This indicates to the software that the line is a comment, in other words, the instructions following the commenting symbol will not be executed, and may help us explain what the programme is doing. This is a very good and recommended practice. Since the commented instructions are not executed, their use enables us to write flexible code as we can comment out any lines that we do not want or that do not need to be executed, without having to delete them. We will come back to this point.

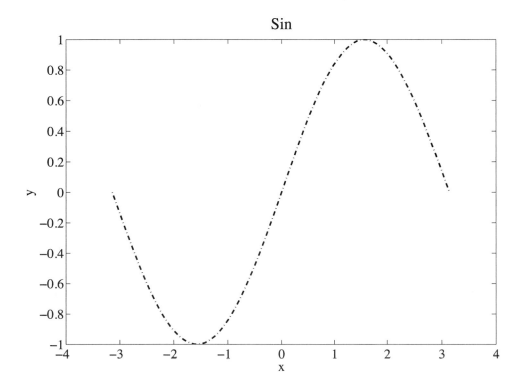

Figure 5.3: Output of the script called my_plot_script.m as shown in the code described in this chapter, with frequency=1 and whichplot=1.

As mentioned by the first comment in the programme above, the clear command deletes any existing variables stored in the workspace. Next we define a variable called frequency whose value can be changed to plot a trigonometric function whose frequency is given by this value. In future use this is a value that can be modified to change the plot.

The variable whichplot will help us decide whether we want to plot the sine or the cosine functions. When whichplot equals one, we will plot the sine; otherwise we plot the co-sine. Please notice that we have commented out a line where whichplot is given a value different from one; and remember that this line will not be executed.

Since a commented line is not executed, we can use this to re-use lines of code at a later stage.

The decision regarding what plot is shown is made with an if statement. When the variable whichplot is equal to 1 we plot the sine function in black with a dash-dot pattern; otherwise we plot the cosine function in black with a dashed line pattern. With the programme written as shown above, we would obtain a plot of the sine function $\sin(x)$ and the result is shown in Figure 5.3. If instead we wanted to plot $\cos(3x)$ we need to change the value of the variable frequency from 1 to 3 and the variable whichplot to 2, or uncomment the appropriate line, and run the programme again. The result of these changes can be seen in Figure 5.4. All we had to do to create these figures was to change a couple of values and re-run the programme; compare this with the prospect of having to re-write the entire series of commands just to change the frequency of the function. As you can imagine, when the number of lines of code becomes larger and larger, the flexibility offered by the scripting capabilities of MATLAB and Octave becomes an essential feature of the language.

The scripting capabilities of MATLAB and Octave are an essential feature of the language and offer great flexibility to us as programmers.

5.5.2   Functions in m-Files

IN THE PREVIOUS SECTION WE have seen how to put commands together in a script. However, it is sometimes useful to write programmes that take an input, carry out a prescribed procedure and return an output.

These scripts are known as *function m-files* and they also have the ".m" extension. Let us consider a function m-file

A *function m-file* is a script that takes an input, executes some operations and returns an output.

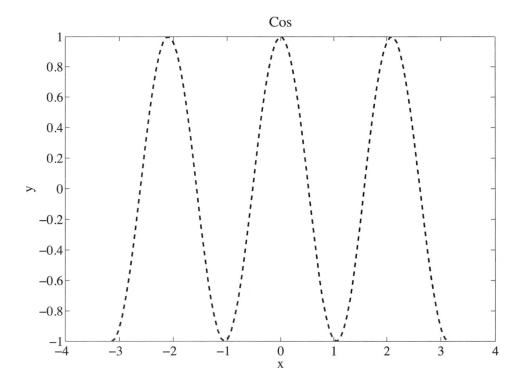

Figure 5.4: Output of the script called `my_plot_script.m.` for `frequency=3` and `whichplot=2`.

called `myfunction.m.` The structure of the function m-file is the following:

```
function [out1,out2,...]=myfunction(in1,in2, ...)
% Comments to be used as help for the function
% Shown when requesting help about the function


    Code to be executed when
    this function is called

out1= ...
out2= ...
end
```

There are several things to note here:

- First of all, the script starts with the word **function**.

  The script starts with the keyword `function`.

- We then have the variables where the outputs of the procedure will be returned (in square brackets), in this case `out1`, `out2`, etc.

  The output variables are placed in square brackets.

- Note that the next character is an equal sign followed by the name of the function. It is important to note that the name of the function must correspond to the name of the script. In this case, since the function is called `myfunction` the script must be saved with the name `myfunction.m`.

  The name of the function and the name of the file must be the same.

- The name of the function is followed by the variables that are used to store the inputs (in round brackets).

  The input variables are placed in round brackets.

- We must assign the values that the function is expected to return.

  The output variables must be assigned a return value.

- Finally, it is important to realise that the input and output variables `in1`, `in2`,... `out1`, `out2`,... are *dummy variables*. In other words, they serve as placeholders within the definition of the function and their scope is valid only inside the function itself.

  The input and output arguments are dummy variables.

The fact that the input and output arguments are dummy variables provides flexibility when writing code as well as at running time. This means that we do not necessarily have to use the variable names used in the definition when calling the function with actual input arguments.

Let us have a look at an example. Imagine that we need to calculate the area and circumference of a circle of a given diameter d. We can do this directly in the terminal. However, we can write a function that can be re-used to calculate these measures for any given circle.

```
function [area, circumference] = my_circle(d)
% This function calculates the area and
% circumference of a circle of diameter d.

% Let us calculate the radius
r = d/2;

% We can use r to calculate the area
area= pi*r^2;

% and the circumference
circumference = 2*pi*r;
end
```

A function that takes a diameter d as an input, and returns the area and circumference of a circle.

We shall save the function in a file called my_circle.m. We can then call the function from the command line in the software, for instance for a circle of diameter 3.

```
> [a,c]=my_circle(3)

a =

    7.0686
c =

    9.4248
```

We can call the function as it is done for any other MATLAB and Octave command.

In this case, the first variable will receive the result for the area, whereas the second one the result for the circumference.

You will notice that after the definition of the function, we have included a couple of comments. These comments are usually used to provide an explanation about what the function does and how to use it. Not only is this help available when looking at the function file, but it is also available in the command line. We can obtain the help provided by the comments at the beginning of the function by using the `help` command followed by the name of the function:

```
> help my_circle

   This function calculates the area and
   circumference of a circle of diameter d.
```

The `help` command can be used to display the explanation added to the function file.

Finally, a word of caution: MATLAB and Octave have a number of built-in functions (see Section 5.6), and you should avoid re-using the same name of any of those existing functions when writing your own. If for whatever reason you end up naming a function with the name of an existing one, remember that MATLAB and Octave apply precedence rules summarised as follows:

1. Before running a function, MATLAB and Octave check if there are any existing variables in the current workspace that match the name given. If so, you cannot run the function until the variable is cleared from the workspace.

   Precedence for the usage of functions.

2. Local functions within the current file take precedence.

3. If no local function is present, MATLAB and Octave check for functions in the current folder.

4. Finally, functions elsewhere on the path are used, in order of appearance.

5.6   Built-In Functions

MATLAB AND OCTAVE INCLUDE A number of ready-made m-files that perform a number of procedures and as such

MATLAB and Octave have a number of functions that are ready to be used.

there is no need to write new files and procedures to carry out those tasks.

There are many of these functions and, as a matter of fact, we have already used a few of them. For more information about their usage, you can type help followed by the name of the function directly in the command line of MATLAB and Octave. Here we provide a list of some of the most common ones:

You can obtain more information about these functions using the help command.

5.6.1   Matrix and Vector Functions

- eye   Identity matrix.

- zeros   Matrix of zeros.

- ones   Matrix of ones.

- diag   Create and extract diagonals.

- triu   Upper triangular matrix.

- tril   Lower triangular matrix.

- linspace   Linearly spaced vector.

- size   Array size.

- length   Length of a vector.

- find   Find indices of nonzero entries.

- sdv   Singular value decomposition.

- inv   Inverse of a matrix.

- reshape   Modify the shape of a matrix.

- det   Determinant of a matrix.

These functions are useful to create and manipulate matrices and vectors.

- rank   Rank of a matrix

- rand   Matrix with random elements.

- magic   Magic square matrix.

- eig   Eigenvectors and eigenvalues of a matrix.

- chol   Cholesky factorisation.

5.6.2   Trigonometric Functions

- sin   Sine.

- sinh   Hyperbolic sine.

- asin   Inverse sine.

- asinh   Inverse hyperbolic sine.

- cos   Cosine.

- cosh   Hyperbolic cosine.

- acos   Inverse cosine.

- acosh   Inverse hyperbolic cosine.

- tan   Tangent.

- tanh   Hyperbolic tangent.

- atan   Inverse tangent.

- atan2   Four quadrant inverse tangent.

- atanh   Inverse hyperbolic tangent.

- sec   Secant.

- sech   Hyperbolic secant.

MATLAB and Octave know about trigonometric functions.

- `asec`   Inverse secant.

- `asech`   Inverse hyperbolic secant.

- `csc`   Cosecant.

- `csch`   Hyperbolic cosecant.

- `acsc`   Inverse cosecant.

- `acsch`   Inverse hyperbolic cosecant.

- `cot`   Cotangent.

- `coth`   Hyperbolic cotangent.

- `acot`   Inverse cotangent.

- `acoth`   Inverse hyperbolic cotangent.

5.6.3   Functions for Complex Numbers

- `abs`   Absolute value.

- `angle`   Phase angle.

- `conj`   Complex conjugate.

- `imag`   Complex imaginary part.

- `real`   Complex real part.

- `unwrap`   Unwrap phase angle.

- `isreal`   True for real array.

- `cplxpair` Sort numbers into complex conjugate pairs

These functions enable us to manipulate complex numbers and functions.

5.6.4 Exponential and Logarithmic Functions

- exp   Exponential.

- log   Natural logarithm.

- log10   Common (base 10) logarithm.

- log2   Base 2 logarithm and dissect floating point number.

- pow2   Base 2 power and scale floating point number.

- sqrt   Square root.

- nextpow2   Next higher power of 2.

These functions let us use exponentials and logarithms.

5.6.5 Rounding and Reminder Functions

- fix   Round towards zero.

- floor   Round towards minus infinity.

- ceil   Round towards plus infinity.

- round   Round towards nearest integer.

- mod   Modulus (signed remainder after division).

- rem   Remainder after division.

- sign   Signum.

The functions listed here provide support for rounding numbers and calculating reminders.

5.6.6   Special Functions

- `airy`   Airy functions.

- `besselj`   Bessel function of the first kind.

- `bessely`   Bessel function of the second kind.

- `besselh`   Bessel functions of the third kind (Hankel function).

- `besseli`   Modified Bessel function of the first kind.

- `besselk`   Modified Bessel function of the second kind.

- `beta`   Beta function.

- `betainc`   Incomplete beta function.

- `betaln`   Logarithm of beta function.

- `ellipj`   Jacobi elliptic functions (MATLAB only).

- `ellipke`   Complete elliptic integral (MATLAB only).

- `erf`   Error function.

- `erfc`   Complementary error function.

- `erfcx`   Scaled complementary error function.

- `erfinv`   Inverse error function.

- `expint`   Exponential integral function (MATLAB only).

- `gamma`   Gamma function.

- `gammainc`   Incomplete gamma function.

- `gammaln`   Logarithm of gamma function.

MATLAB and Octave know a number of very useful special functions.

- `legendre`   Associated Legendre function.

- `cross`   Vector cross product.

5.6.7   Number Theoretic Functions

- `factor`   Prime factors.

- `isprime`   True for prime numbers.

- `primes`   Generate list of prime numbers.

- `gcd`   Greatest common divisor.

- `lcm`   Least common multiple.

- `rat`   Rational approximation.

- `rats`   Rational output.

- `perms`   All possible permutations.

- `nchoosek`   All combinations of n elements taken k at a time.

MATLAB and Octave provide support for number theory.

5.6.8   Coordinate Transformations

- `cart2sph`   Transform Cartesian to spherical coordinates.

- `cart2pol`   Transform Cartesian to polar coordinates.

- `pol2cart`   Transform polar to Cartesian coordinates.

- `sph2cart`   Transform spherical to Cartesian coordinates.

It is possible to carry out basic coordinate transformations with MATLAB and Octave.

- hsv2rgb   Convert hue-saturation-value colours to red-green-blue (MATLAB only).

- rgb2hsv   Convert red-green-blue colours to hue-saturation-value (MATLAB only).

5.6.9   Statistics

- mean   Arithmetic mean.

- median   Median.

- cov   Variance and covariance.

- corrcoef   Correlation.

- sum   Sum of elements.

- prod   Product of elements.

- hist   Histogram.

- max   Largest element.

- min   Smallest element.

Some statistics functions available in MATLAB and Octave.

5.6.10   Data Interpolation

- pchip   Piecewise cubic Hermite interpolation.

- interp1   1D interpolation.

- interp1q   1D interpolation without error checking.

- interp2   2D interpolation.

Functions to deal with data interpolation and spline interpolation.

- `interp2`   3D interpolation.

- `interpn`   N-dimensional interpolation.

- `griddata`   Regular mesh from irregular data using interpolation.

- `spline`   Cubic spline interpolation.

5.6.11   Polynomials

- `roots`   Find the roots of a polynomial.

- `poly`   Convert roots to a polynomial.

- `polyval`   Polynomial evaluation.

- `polyvalm`   Polynomial evaluation with a matrix argument.

- `residue`   Partial fraction expansion.

- `polyfit`   Fit polynomial to data.

- `polyder`   Polynomial differentiation.

- `polyint`   Polynomial integration.

- `conv`   Convolution.

- `deconv`   Deconvolution.

Functions to deal with polynomials and fitting.

5.6.12   Finite Differences

- diff   Approximate derivative.

- gradient   Approximate gradient.

- del2   Discrete Laplacian.

MATLAB and Octave know how to calculate finite differences.

5.6.13   Differential Equations

- ode45   Runge-Kutta Method (only MATLAB).

- lsode   Ordinary differential equation solver (only Octave).

MATLAB and Octave have many other functions to deal with ODEs; please check the help.

5.6.14   Optimisation and Root Finding

- fminbnd   Scalar bounded nonlinear function minimisation.

- fzero   Scalar nonlinear zero finding.

Optimisation and root finding functions.

5.6.15   Fourier Transforms

- fft   Fast Fourier transform.

- fft2   2D fast Fourier transform.

- fftn   N-dimensional fast Fourier transform.

Functions to deal with Fourier transforms.

- `ifft`   Inverse fast Fourier transform.

- `ifft2`   Inverse 2D fast Fourier transform.

- `ifftn`   Inverse N-dimensional fast Fourier transform.

- `fftshift`   Shift zero-frequency to the centre of spectrum.

- `ifftshift` Inverse `fftshift`.

5.7   *Function Handles*

IN THIS CHAPTER WE HAVE seen how to construct and programme a function in MATLAB and Octave. Although these functions can be used with good effect to build more complex programmes, sometimes it is necessary to exploit further capabilities offered by the software.

For example, from time to time when a process requires that a function calls another function, it may be more straightforward to call the second function indirectly. In order to do that we need to use a *function handle,* which allows us to do various things such as calling a function from another function, creating functions of functions, and even storing them in data structures that can be recalled later.

A function handle lets us manipulate a function from another function.

In order to build a function handle we simply need to put the @ symbol before the name of the function; for instance if we have a function called `MyFunctionName`, a handle `fhandle` for this function is defined as follows:

```
fhandle = @ MyFunctionName
```

A function handle is denoted by the use of the @ symbol.

The software will map the handle to the specified function and this information is stored in the handle itself. It is important to note that a handle will keep this mapping even when the function is out of scope. For example, if during the execution of a programme the path is changed and another function with the same name has a higher priority, the handle will still call the original function stored in the older path. Please refer to the last part of Section 5.5.2 for a brief discussion of function precedence.

A function handle has a persistent mapping to the function it refers to, regardless of function precedence rules.

In the following sections we will discuss some applications of function handles.

5.7.1   Anonymous Functions

HANDLES CAN BE USED VERY effectively when we need to define a function on the fly, in other words, a procedure that is simple enough to be defined in a programme without the need of creating a separate m-file. These type of functions are called *anonymous functions*. The syntax of an anonymous function is as follows:

An anonymous function is simple enough to be defined in one line of code.

```
function_name = @(var) expression with var
```

where var is the variable used in the function.

For example, if we wanted to calculate repeatedly the value of the following expression:

$$f(x) = x^3 + 3x - 1, \qquad (5.12)$$

we would like to do this for various values of x, and thus we can certainly write the full expression every time, or create a function in a separate m-file. However, this can be simplified greatly with an anonymous function defined as follows:

An anonymous function can be used for simple processes that do not require a full m-file of their own.

```
> CUBIC = @(x) x.^3+3*x-1;
```

We have defined a function called CUBIC where the input
is given by the variable x. We know this because x is the
variable enclosed in the expression @(). This is followed by
the expression given in Equation (5.12). All we have to do to
use this function is simply call it as we have been doing for
other functions:

```
> CUBIC(1)

ans =

      3

> CUBIC(2)

ans =

     13
```

An anonymous function can be
executed in the same way other
functions are.

5.7.2   Arrays of Function Handles

AS WE MENTIONED BEFORE, IT is possible to store function handles as we do any other variable. We can think of
function handles as objects that point to other functions and
facilitate their use without using complicated manipulations.
Let us take a look at the following example. We can create
a function handle for any built-in function; for instance
we can do so for the sine or any of the other trigonometric
functions:

```
> sine_handle = @sin;
```

As such, we can use the new handle as a synonym for the
function:

```
> sine_handle(pi/2)

ans =

      1
```

We can think of a function handle as a flexible synonym for the original function.

This particular example is not very practical, but it helps us understand how function handles behave. With this in mind, we can think of creating new structures that refer to the original functions themselves. For instance, let us imagine that we want to create a flexible piece of code that enables us to select any of the main trigonometric functions sin, cos and tan and generate a plot for the function we chose. We can do this in a script as follows:

```
trigFun = {@sin, @cos, @tan};

plot (trigFun{2}(-pi:0.01:pi));
```

Function handles can be stored in arrays that can be later retrieved for use.

Here we have created an array of function handles to the trigonometric functions defined in the software. In the second line we are calling the array of function handles with the argument 2. In this case, the handle will use the second entry of the array, i.e. it will evaluate the cos function and plot it for the interval $[-\pi, \pi]$. If we want to evaluate any of the other functions then we just simply change the argument to the appropriate element in the array.

5.7.3   Function Handles as Arguments

AS WE HAVE SEEN ABOVE, a function handle in MATLAB and Octave is a flexible way we have to call a function indirectly. We have mentioned as well that we can pass

Function handles can be used as arguments to pass to other functions.

function handles to other functions, and this offers a way of evaluating quantities of interest.

For example, let us consider the problem of finding the minimum of the following function:

$$y(x) = 3x^3 - 9x + 5. \qquad (5.13)$$

We can certainly do this by hand but let us see how this can be done with MATLAB and Octave. From Section 5.6.14 we know that the built-in function fminbnd is able to minimise a function. Let us see how this can be done with the use of a function handle.

The function fminbnd can be used to find the minimum of a function.

First let us define an anonymous function for the expression in Equation (5.13):

```
> f=@(x) 3*x.^3-9*x+5

f =

      @(x) 3 * x .^ 3 - 9 * x + 5
```

We could indeed have created an m-file for this function; this is left for the reader as an exercise. We can now use this function handle to calculate the minimum by passing it as an argument to the fminbnd command as follows:

```
> x_min=fminbnd(f,0.5,2)

x_min =

      1.0000
```

We can pass the function handle created above to the fminbnd function to find the minimum.

The first argument given to fminbnd is the function handle f created above; the next two values correspond to the interval where the minimum is being sought. It can be easily verified that the value returned by fminbnd is indeed the minimum for the function in Equation (5.13).

5.8   Debugging

IN THE DEVELOPMENT OF ANY script there may be occasions where things are not working quite as one expects. The process of finding and minimising the number of these so-called *bugs* is known as debugging. In order to assist with the tasks such as setting breakpoints, line-by-line execution or examining values, MATLAB and Octave provide some very useful commands.

Debugging tools enable us to stop the execution of a programme to get rid of bugs in a programme.

MATLAB's graphical editor enables the use of debugging tools graphically. For further information about MATLAB graphical interface consult The MathWorks[1]. In the following discussion we will concentrate on commands that are typed either in the command-line or in the scripts themselves as these are available in Octave, too. Please note that although the commands described here have the same name, they have sometimes different syntax.

[1] The MathWorks - http://www.mathworks.com

ONE EFFECTIVE WAY TO START debugging a script or programme is by setting breakpoints in the code, in other words, places where the execution of the programme is paused allowing us to examine the values of variables and the logic of the code. This can be achieved with the command dbstop:

```
dbstop in filename at linenumber
```

MATLAB

```
dbstop('filename', linenumber )
```

Octave

The software will stop at the linenumber specified and will display, in the command-line, the next line to be executed.

In order to let the user know that the debug mode is on, the prompt will look different. In the case of MATLAB, the letter K is used at the beginning of the prompt as shown below:

```
K>>
```

MATLAB

In Octave, the word debug is used as follows:

```
debug>
```

Octave

Once the debug mode is on, the software waits until the user provides input to continue the execution of the programme. In order to execute the script one line at a time the command

```
> dbstep
```

dbstep executed the programme one line at a time.

can be used. If instead we want to resume the normal execution of the script, we can use the command

```
> dbcont
```

dbcont resumes the execution of the programme.

We can leave the debug mode using the command

```
> dbquit
```

dbquit quits the debug mode.

LET US CONSIDER THE FOLLOWING function as an example to debug:

```
function result = debuggy(x)
    n=length(x);
    result=n/x;
end
```

We can call this function with an input given by a vector whose elements are the numbers from 1 to 10 and request for the function to stop in line 1:

```
> x=[1:10];
> dbstop in debuggy at 1
> y = debuggy(x)
```
MATLAB

```
> x=[1:10];
> dbstop('debuggy', 1)
> y = debuggy(x)
```
Octave

which results in the execution of the programme at the desired location:

```
2    n=length(x);
K>>
```
MATLAB

```
stopped in debuggy.m at line 2
2: n=length(x);
debug>
```
Octave

We can move to the next line of code as follows:

```
K>> dbstep
3    result=n/x;
```
MATLAB

```
debug> dbstep
stopped in debuggy.m at line 3
3: result=n/x;
```
Octave

As we move to the next line we obtain the following error:

```
K>> dbstep
Error using  /
Matrix dimensions must agree.

Error in debuggy (line 3)
result=n/x;
```
MATLAB

```
debug> dbstep
error: debuggy: operator /:
nonconformant arguments (op1 is 1x1, op2 is 1x10)

error: called from debuggy.m at line 3, column 7
```
Octave

In the simple example above we can see that the debugging tools tell us that the line that is causing the error is the execution of line 3 of the function called debuggy.m. The error is caused because we are trying to divide a scalar by a vector. This can be solved by changing line 3 to use the dot-division operator:

```
result = n./x;
```

This is one potential solution, and it may be the case that this is not the behaviour we require from the function. It is therefore highly recommended to review the logic and expected results of any script you write.

Once one or many breakpoints have been set up with dbstop command, it is possible to list them all with the command

```
> dbstatus
```

dbstatus lists all the breakpoints currently set.

and finally, we can clear all breakpoints with the command

```
> dbclear in filename
```

MATLAB

```
> dbclear('filename')
```

Octave

Finally, we would like to mention another useful command for debugging:

```
> keyboard
```

keyboard stops the execution of a programme and gives control to the keyboard.

This command stops the execution of a script and the control is passed to the user who can interact with the software via the keyboard, making it possible for us to view the values of variables. Normal execution can be regained with the command.

```
> return
```

5.9   Timing

NOW THAT WE HAVE COVERED some of the basics of scripting and programming in MATLAB and Octave, it becomes natural to ask about the performance of a given calculation. The software has a couple of functions that allow us to time the execution of a certain part of the code: tic and toc.

The commands tic and toc help us time the execution of our programmes.

We can think of the combination of these two functions as a stopwatch: tic starts it while toc stops it. The function toc by itself returns the elapsed time in seconds since `tic` was used.

For example, we can check how long it would take our computer to calculate 1000 values of the anonymous function `CUBIC` defined in Section 5.7.1.

```
> tic, for j=1:1000, CUBIC(j); end, toc


Elapsed time is 0.005134 seconds.
```

The combination of `tic` and `toc` provides us with information about how long MATLAB and Octave take in performing a number of calculations.

Please note that the elapsed time will depend on the machine where the code is run and the time quoted above may differ from the one you may obtain.

5.10   Reading and Writing Files

IT IS GREAT TO BE able to use MATLAB and Octave as a powerful calculator and programming environment, but in general there is the need for input and output of large datasets whose direct handling with the keyboard is simply impractical. In cases like these the input and output can be handled more efficiently by using files.

Large datasets for input and output are better handled by files.

MATLAB and Octave are able to deal with formatted and unformatted data. Formatted data use format strings to define exactly how and in what positions of a record the data are stored. Unformatted data require only to specify the number format.

We can use the function `fopen` to open a file for access; the syntax of this function is as follows:

```
fID = fopen(filename, permissions)
```

The command fopen enables us to open a file.

where fID is a file identifier, filename is the name of the file to be opened and permissions lets the software know what it is able to do with the file. A list of possible permissions to be used appears in Table 5.3. A more complete list can be obtained with the help command in the software.

Permission	Meaning
'r'	open file for reading
'w'	open file for writing
'a'	open file for appending
'r+'	open (do not create) file for reading and writing
'w+'	open or create file for reading and writing
'a+'	open or create file for reading and writing, append data at the end

Table 5.3: Permissions for opening files with fopen.

5.10.1 Formatted Files

MATLAB AND OCTAVE ARE ABLE to read and write formatted text files; both operations act on matrix elements on a column-by-column basis, however the reading of text files is done row-by-row which means that sometimes a transposition is needed.

Formatted data require the use of format strings to define how and where in a record the data are stored.

5.10.2 Reading Formatted Files

LET US CONSIDER SOME DATA that are stored in a table formatted in two columns and saved in a plain text file called myData.txt as follows:

```
10    2340
20    3450
30    4580
40    5960
50    6788
60    7890
...
```

The example data stored in myData.txt are stored in two columns in plain text.

We know that we can open the file using fopen, but to read the contents of the file we have to make use of another function: fscanf.

This function reads the data from a file with a particular identifier and converts it according to a format specified by the user. The syntax of this function is as follows:

```
[A,count] = fscanf(fID,format,size)
```

We can read formatted data from an open file with the fscanf command.

Array A will hold the contents of the file and count is the number of elements that are read, whereas fID is the file identifier defined when the file was opened with fopen; the argument size is optional and it puts a limit on the number of elements that can be read.

Finally, the string format specifies the way in which data should be converted. This string starts with the symbol % followed by a specifier. Possible specifiers are listed in Table 5.4.

The format specifier can also contain sub-specifiers, namely, asterisk (*) and width. These are optional and can be described as follows:

- * - An asterisk indicates that the data are to be read from the stream but ignored (i.e. they are not stored in the location pointed by an argument).

Optional sub-specifiers for reading formatted data with fscanf.

- width - Specifies the maximum number of characters to be read in the current reading operation.

Specifier	Meaning
'%d'	Signed integer, base 10
'%i'	Signed integer, base determined from values
'%u'	Unsigned integer, base 10
'%f'	Floating point number
'%s'	String of characters

Table 5.4: Format definitions for reading and writing data.

With this information we can now understand the way to read the example file we mentioned before:

```
> fID = fopen('myData.txt','r');
> a = fscanf(fID,'%2d%4d', [2 inf]);
> fclose(fID);
> a=a';
```

Full code to read the content of a file with formatted data.

Let us explain what the code above is doing in each line:

1. The meaning of the first line must be clear from the description in Section 5.10: it simply opens the file to be read.

2. The second line instructs the software to go through the file and read the first two characters which will be stored as an integer, then read the next four characters and store them as integers, too.

The first three lines of the code above look simple, but they do very powerful things to deal with files.

3. The third line should be very straightforward to understand: fclose simply instructs the software to close the file once the reading process has finished. This is needed to avoid undesired changes to the file.

fclose closes the file opened with fopen.

Notice that we have instructed the software to store the data in a matrix of 2 by inf; this is because we are not certain about the total number of entries. Finally we must transpose the matrix in order to get the same format as stored in the data file.

5.10.3   Writing Formatted Files

NOT ONLY IS IT IMPORTANT to read files with the commands explained above but also to store the result of calculations carried out by MATLAB and Octave. This can be done in a plain text file that can then be stored and possibly read and processed later on. We can achieve this task by using the command fprintf. The syntax of this function is as follows:

```
count = fprintf(fID,format,A,...)
```

The command fprintf lets us write formatted data to an open file.

where count is the number of elements written, fID is the identifier of the file where the data will be stored and format specifies the format of the information that will be written, following the syntax explained in Section 5.10.2. Finally the array(s) A, ... are the matrices whose elements will be written in the file specified by fID.

Note that the use of fprintf assumes that the destination file has been opened and that the file is ready to store the data specified.

The use of fprintf assumes that the destination file has been opened.

Let us consider an example: we will create a file that stores a simple table of some values for the exponential function

$$y(x) = \exp(x). \qquad (5.14)$$

We can achieve this task as follows:

```
> x=0:0.1:1;
> y=[x;exp(x)];
> fid=fopen('exp.txt','w');
> fprintf(fid,'%6.2f %12.8f\n',y);
> fclose(fid);
```

We can generate a file with a tabulation of the exponential function with a combination of fopen and fprintf.

In the first line we have created a vector with 11 values for the independent variable x (a sequence from 0 to 1 in steps of 0.1). We then create an array y whose first column is the value of x and the second column is the value of $\exp(x)$. The third line instructs the software to open a file called exp.txt; notice that we have passed the argument 'w' to the fopen command which implies that the software has writing permissions. The fourth line stores the values of the matrix y in two columns. Notice that the formatting specifiers end with the string \n which forces the file to create a new line and thus the next set of values will be printed in the next row in the file. Instructions such as \n at the end of the formatting portion of the command are called *control characters* and Table 5.5 shows some of the most common ones.

Remember to enable writing permissions when creating files for storing data.

Control characters such as '\n' enable us to format the data that will be stored in the file.

Specifier	Meaning
'\n'	New line
'\r'	Beginning of new line
'\b'	Backspace
'\t'	Tab
'\f'	New page

Table 5.5: Control characters used in formatting output.

Finally, the fifth line closes the file. The result of these instructions creates a file called exp.txt and its contents are as follows:

```
0.00    1.00000000
0.10    1.10517092
0.20    1.22140276
0.30    1.34985881
0.40    1.49182470
0.50    1.64872127
0.60    1.82211880
0.70    2.01375271
0.80    2.22554093
0.90    2.45960311
1.00    2.71828183
```

These are the contents of the file exp.txt generated with the code above.

5.10.4 Binary Files

IN THE PREVIOUS SECTION WE dealt with formatted files, but it may be the case that it is necessary for us to work with binary files directly. Both types of files may seem very similar but it is important to understand that they encode data in different ways: the bits in a formatted file represent characters, whereas the bits in a binary file represent custom data. MATLAB and Octave are capable of dealing with binary files, too.

Binary files contain custom data that are not formatted.

5.10.5 Writing Binary Files

IN ORDER TO WRITE A binary file, we need to use the command fwrite, whose syntax is as follows:

```
fwrite(fID, mtx, precision)
```

The command fwrite lets us write binary data to a file.

where fID is the identifier of the file to store the data and mtx is the matrix to be stored. Finally, precision refers to the type of data that is being written. Table 5.6 shows some of the specifiers that can be used for the precision of the data.

Specifier	Meaning
'uint'	Integer, unsigned 32 bits
'uint8'	Integer, unsigned 8 bits
'int	Integer, signed 32 bits
'int8	Integer, signed 8 bits
'double'	Floating, 64 bits
'char'	Character

Table 5.6: Precision specifiers for reading and writing binary data.

Let us imagine that we are interested in storing the magic matrix given by magic(3) in a file called magic3.dat. We can achieve this with the following commands:

```
> fid = fopen('magic3.dat', 'w');
> fwrite(fid, magic(3));
> fclose(fid);
```

We can write binary data with a combination of the fopen and fwrite commands.

The first line opens up the file for writing. We then instruct the software to write the content of the 3×3 magic matrix and close the file.

5.10.6 Reading Binary Files

BY NOW WE CAN DEFINITELY start seeing a pattern: we can read a binary file with a command similar to those we have seen above. In this case, the reading is done with the command fread:

```
[A, c] = fread(fID, mtx, precision)
```

The command fread enables us to read binary files.

where fID is the identifier of the file to be read into matrix A; mtx is an $m \times n$ matrix that will be read column by column

out of the file with identifier fID and the precision of the
data can be specified using strings shown in Table 5.6.

For instance if we wanted to read the content of the binary
file we created in Section 5.10.5 called magic3.dat and
which contains three columns and three rows, we can use
the following commands:

```
> fid = fopen('magic3.dat','r');
> A = fread(fid, [3, 3], '*uint8');
> fclose(fid);
```

We can read the contents of a
binary file with the help of the
commands fopen and fread.

The * in the precision specification used in the example
above means that the output has the same class as that
in the input. Notice that we first need to open the file for
reading by using fopen in conjunction with the 'r' flag. We
can then read the contents with the fread command and
finally we close the file with the help of fclose.

5.11 Summary

IN THIS CHAPTER WE HAVE brought together the concepts
and techniques that we have been discussing in the first
four chapters of this book. We have also added some other
tools that improve and enhance our experience with both
MATLAB and Octave.

We have gone from using MATLAB and Octave as elaborate
calculators, to a fully flexible programming environment
based on m-files. These files are plain text files that contain
lines of code which are executed by MATLAB and Octave
one by one. We now know how to provide a coherent logical flow to our programmes using relational and logical

operators. These together with loops and conditional statements enable us to create elaborate procedures that the software can execute.

Furthermore, we have also seen how functions can give us the power of carrying out processes in a reproducible way, with the added advantage of carrying out the calculations with different parameters (inputs) and obtaining the corresponding answers (outputs). In many cases, MATLAB and Octave already have a number of built-in functions that can be readily used. Finally, we also discussed the use of function handles as arguments for other functions.

In the next chapter we will provide a small selection of examples that showcase some of the capabilities of MATLAB and Octave.

5.12   Exercises

1. Write a function that takes three parameters a, b and c to solve a quadratic equation $ax^2 + bx + c = 0$. Your function should calculate the solutions based on the value of the discriminant $D = b^2 - 4ac$ to let the user know if the roots are both real or imaginary, or if there is only one solution. Also check if the equation is indeed quadratic; otherwise return nan.

2. Create anonymous functions for the following expressions and plot them as requested:

   (a) Calculate
   $$f(x) = 2x^3 - 3x^2 + x - 5,$$
   and plot it for $x \in [-2, 4]$.

   (b) Calculate
   $$g(t) = \exp(-0.2t)(2\cos(t-1)),$$
   and plot it in polar coordinates for $t \in [0, 4\pi]$.

   (c) Calculate
   $$h(y) = y^4 - 4y + 5,$$
   and plot it for $y \in [-10, 10]$.

3. Recursive functions are functions that call themselves either directly of indirectly. Write a recursive function that returns the n-th term of the Fibonacci series $1, 1, 2, 3, \ldots, a_{n-2}, a_{n_1}, a_n$, where $a_n = a_{n-2} + a_{n-1}$.

4. Write a function using a for loop to calculate the factorial of a number and calculate the time it takes to produce 10! and 15!. Compare your results and performance to a function that uses recursion to do the same task. Also compare them to the built-in factorial function.

5. Using logical operators create a script that plots the following function:

$$g(x) = \begin{cases} x, & -5 \le x < 0, \\ \sin(x), & 0 \le x \le 5. \end{cases}$$

6. Plot the following function

$$y = \frac{\sin x}{x}$$

for x=-1:0.1:1. Is there something wrong with the plot? Try sorting the problem using the eps value defined in the software. After solving the issue, produce a plot in the interval $[-10, 10]$.

7. Write a script using a while loop that calculates the successive partial sums

$$1 + \frac{1}{2} + \frac{1}{3} + \cdots + \frac{1}{n}$$

until successive sums are within 0.01 of each other, printing on every step the result of the sum.

8. Using a while loop determine how many terms are needed for successive sums of 2^k to exceed 1000.

9. Write a function that takes the coefficients of a cubic function

$$y(x) = ax^3 + bx^2 + cx + d$$

and the interval $[x_1, x_2]$ to create a plot. Your script should also create a file with the values of x and y.

10. The bisection method is an algorithm to find the root of a function in a given interval. The method cuts the interval in two portions and checks which one of them contains the root by checking for changes of sign in the values of the function. Pseudo-code for an implementation of the

method is given below. Use the pseudo-code to create a
MATLAB and Octave function for the method.

```
while (interval [a,b] is not "very small")
{
    m=(a+b)/2 % mid-point
    if (sign of f(m) is different of sign of f(b))
    { use interval [m,b] in the next iteration
      (i.e. replace a with m) }
    else
    { use interval [a,m] in the next iteration
      (i.e. replace b with m) }
}
approximate root is (a+b)/2
```

We recommend that you find out what the function feval
does and use it in your implementation. Finally, use your
bisection function to calculate the root of

$$f(x) = \exp(-x)\,(3.2\sin(x) - 0.5\cos(x))$$

in the interval $[3, 4]$.

6
MATLAB® and Octave in Action

IN THE PREVIOUS FIVE CHAPTERS we have discussed a wide range of topics, from what we called "the essential essentials" of MATLAB and Octave to the usage of various different objects that the software provides. We have also highlighted the flexibility that is provided by the scripting language used by MATLAB and Octave.

In this chapter we present a few examples that showcase the use of both MATLAB and Octave in context. This should not be taken as a thorough and rigorous discussion of the topics used, but rather as a small smörgåsbord that represents the handling of some of the techniques discussed earlier on in the book. We therefore recommend reading this chapter with the help of a suitable textbook, which may help clarify some of the concepts and ideas, as well as a running instance of MATLAB or Octave. We have provided some basic references in each example with the intention of guiding you to find further information on each of the topics addressed. Finally, we would like to encourage you to try the examples presented here on your own computer; feel free to adapt the scripts and methods to your own needs.

The topics and examples presented aim to show the usage of MATLAB and Octave within context.

Try the examples on your own computer and start implementing your own.

6.1   Linear Algebra: Linear Combinations

GIVEN THAT THE MOST BASIC object in MATLAB and Octave is a matrix, it seems natural to use the software for linear algebra applications first. Consider the following four-dimensional vectors:

Vector and matrix manipulations are a natural application for the software.

$$\mathbf{v_1} = (2, 0, -1, 7), \qquad (6.1)$$

$$\mathbf{v_2} = (-1, 3, 4, 7), \qquad (6.2)$$

$$\mathbf{v_3} = (0, 2, 5, 7), \qquad (6.3)$$

$$\mathbf{v_4} = (0, 1, 3, 5). \qquad (6.4)$$

We would like to find out if the vector

$$\mathbf{v} = (5, 4, 12, 33) \qquad (6.5)$$

is a linear combination[1] of the vectors $\mathbf{v_1}$, $\mathbf{v_2}$, $\mathbf{v_3}$ and $\mathbf{v_4}$.

[1] Strang, G. (2003). *Introduction to Linear Algebra*. Wellesley-Cambridge Press

In order to tackle this problem we need to form the expression

$$\sum_{i=1}^{4} c_i \mathbf{v_i} = \mathbf{v} \qquad (6.6)$$

and solve the corresponding linear system to find the coefficients c_i:

$$c_1(2, 0, -1, 7) + c_2(-1, 3, 4, 7)$$
$$+ c_3(0, 2, 5, 7) + c_4(0, 1, 3, 5) = (5, 4, 12, 33). \quad (6.7)$$

This leads us to the following linear system:

$$2c_1 - c_2 = 5, \qquad (6.8)$$

$$3c_2 + 2c_3 + c_4 = 4, \qquad (6.9)$$

$$-c_1 + 4c_2 + 5c_3 + 3c_4 = 12, \qquad (6.10)$$

$$7c_1 + 7c_2 + 7c_3 + 5c_4 = 33. \qquad (6.11)$$

The linear combination leads us to a linear system of equations.

We can write the system of equations in terms of matrices as

$$\mathbf{Ac} = \mathbf{b}, \qquad (6.12)$$

which means that the system to solve is given by

$$\begin{pmatrix} 2 & -1 & 0 & 0 \\ 0 & 3 & 2 & 1 \\ -1 & 4 & 5 & 3 \\ 7 & 7 & 7 & 5 \end{pmatrix} \begin{pmatrix} c_1 \\ c_2 \\ c_3 \\ c_4 \end{pmatrix} = \begin{pmatrix} 5 \\ 4 \\ 12 \\ 33 \end{pmatrix}. \qquad (6.13)$$

A linear system can be expressed in terms of matrices.

Notice that matrix $\mathbf{A}$ is made out of the transpose of the original vectors. We solve this problem in MATLAB and Octave with the following script:

```
% This script is saved with the
% name Ch6_LinearAlgebra1.m

clear;
v_1=[2, 0, -1, 7];
v_2 =[-1, 3, 4, 7];
v_3=[0, 2, 5, 7];
v_4=[0, 1, 3, 5];

% Constructing the matrix A
A=[v_1' v_2' v_3' v_4']

% Vector b
b=[5, 4, 12, 33]

% Solving the linear system
c=A\b'
```

We first clear the workspace and define the vectors to work with.

Concatenation lets us construct the matrix.

We then use left-division operator to solve the system. See Section 3.10.

We now invoke the script as follows:

```
> Ch6_LinearAlgebra1

c =

   2.00000
  -1.00000
   3.00000
   1.00000
```

The solution of the linear system is returned by MATLAB and Octave.

It follows that the system is consistent with the solution given by

$$c_1 = 2, \tag{6.14}$$

$$c_2 = -1, \tag{6.15}$$

$$c_3 = 3, \tag{6.16}$$

$$c_4 = 1, \tag{6.17}$$

These non-zero values tell us that the vector $\mathbf{v}$ is a linear combination of the vectors in question.

and hence $\mathbf{v}$ is a linear combination of $\mathbf{v}_1$, $\mathbf{v}_2$, $\mathbf{v}_3$ and $\mathbf{v}_4$ with

$$2\mathbf{v}_1 - \mathbf{v}_2 + 3\mathbf{v}_3 + \mathbf{v}_4 = \mathbf{v}. \tag{6.18}$$

6.2   Linear Algebra: Eigenvalues and Eigenvectors

LET US CONSIDER THE FOLLOWING 3×3 matrix:

$$\mathbf{A} = \begin{pmatrix} 3 & 4 & 8 \\ 1 & -3 & 2 \\ 42 & 15 & -4 \end{pmatrix}; \tag{6.19}$$

we would like to find its eigenvalues and eigenvectors[2]. We need to find the values λ that satisfy the characteristic equation of the matrix $\mathbf{A}$:

[2] Strang, G. (2003). *Introduction to Linear Algebra*. Wellesley-Cambridge Press

$$\det(\mathbf{A} - \lambda\mathbf{I}) = 0, \tag{6.20}$$

where $\mathbf{I}$ is a 3×3 identity matrix. Let us first form the matrix $\mathbf{A} - \lambda \mathbf{I}$:

$$\mathbf{A} - \lambda \mathbf{I} = \begin{pmatrix} 3 & 4 & 8 \\ 1 & -3 & 2 \\ 42 & 15 & -4 \end{pmatrix} - \begin{pmatrix} \lambda & 0 & 0 \\ 0 & \lambda & 0 \\ 0 & 0 & \lambda \end{pmatrix},$$

$$= \begin{pmatrix} 3 - \lambda & 4 & 8 \\ 1 & -3 - \lambda & 2 \\ 42 & 15 & -4 - \lambda \end{pmatrix}, \qquad (6.21)$$

The eigenvalue problem relies on the construction of this matrix.

and then calculate its determinant. Using MATLAB and Octave, we can do all this by using the eig command.

The built-in function eig facilitates this calculation. For other functions see Section 5.6.1.

Finally, let us recall that for each eigenvalue λ we have

$$(\mathbf{A} - \lambda \mathbf{I}) \, \mathbf{x} = 0, \qquad (6.22)$$

where $\mathbf{x}$ is the eigenvector associated with the eigenvalue λ.

We can solve this problem with the following script:

```
% This script is saved with the
% name Ch6_LinearAlgebra2.m

clear;
% Defining the matrix A
A=[3 4 8; ...
   1 -3  2; ...
   42 15 -4];

% Using the function eig
[V lambda] = eig(A)
```

We clear the workspace and define the matrix A.

eig can return two outputs, the eigenvectors V and the eigenvalues lambda.

We can execute the script as follows:

```
> Ch6_LinearAlgebra2

V =

  -0.456435   -0.317100    0.331620
  -0.099460   -0.094505   -0.925529
  -0.884180    0.943672    0.182821


lambda =

Diagonal Matrix
   19.3688           0           0
        0    -19.6154           0
        0           0     -3.7534
```

The eigenvectors are given by the columns of the matrix V.

The eigenvalues are given by the diagonal of the matrix lambda.

The command eig returns a diagonal matrix, lambda, that contains the eigenvalues, and a matrix V whose columns correspond to the eigenvectors. This means that the eigenvalues of matrix **A** in Equation (6.19) are

$$\lambda_1 = 19.3688, \tag{6.23}$$

$$\lambda_2 = -19.6154, \tag{6.24}$$

$$\lambda_3 = -3.7534, \tag{6.25}$$

These are the eigenvalues of matrix **A**.

and the eigenvectors are

$$\mathbf{v_1} = \begin{pmatrix} -0.456435 \\ -0.099460 \\ -0.884180 \end{pmatrix}, \tag{6.26}$$

$$\mathbf{v_2} = \begin{pmatrix} -0.456435 \\ -0.094505 \\ 0.943672 \end{pmatrix}, \tag{6.27}$$

These are the eigenvectors of matrix **A**.

$$\mathbf{v_3} = \begin{pmatrix} 0.331620 \\ -0.925529 \\ 0.182821 \end{pmatrix}. \tag{6.28}$$

With the aid of Equation (6.22) we can verify that the results given by MATLAB and Octave are correct. In this case, we can recast the equation in terms of the eigenvector $\mathbf{v_1}$ to verify that $\mathbf{A}\mathbf{v_1} = \lambda_1\mathbf{v_1}$:

```
> v1= V(:,1)
v1 =

      -0.456435
      -0.099460
      -0.884180


> lambda1=lambda(1,1)
lambda1 =
       19.369


> a = A*v1
a =

       -8.8406
       -1.9264
      -17.1255


> b= lambda1*v1
b =

       -8.8406
       -1.9264
      -17.1255


> a-b
ans =
        0.0000
        0.0000
        0.0000
```

We can use the colon notation to extract the first eigenvector. See Section 3.7.

We can do the same to extract the corresponding eigenvalue.

We can use matrix multiplication to calculate the desired results. See Section 3.8.

Finally, we can verify that the two operations are equal to each other.

Please note that the formatting used above shows the difference a−b as zeros, but depending on the format and precision used you may see a different result but very close to zero. Finally, similar relationships can be verified for the other eigenvectors and eigenvalues, i.e. $\mathbf{A}\mathbf{v_2} = \lambda_2 \mathbf{v_2}$ and $\mathbf{A}\mathbf{v_3} = \lambda_3 \mathbf{v_3}$.

6.3   Portfolio Risk: Minimum Variance and Target Portfolios

IN THIS EXAMPLE WE WILL consider a financial portfolio comprising three assets. In general terms, a portfolio is a collection of assets, and in turn assets are understood to be resources with an economic value and which are held with the expectation that they will provide future benefit or return. Some examples of assets include stocks, bonds, funds or even cash. A couple of important measures of a portfolio are the variance-covariance matrix (Σ) of the assets, as well as their expected returns. The latter can be arranged into a column-vector for easier manipulation ($\bar{\mathbf{R}}$). The risk associated with a financial portfolio is related to its standard deviation, given the assets it contains. Investors are thus interested in balancing the risk of a portfolio and its expected return[3].

[3] Elton, E., M. Gruber, S. Brown, and W. Goetzmann (2009). *Modern Portfolio Theory and Investment Analysis*. John Wiley & Sons

For the purposes of this example we will use the following variance-covariance matrix:

$$\Sigma = \begin{pmatrix} 0.3 & 0.02 & -0.05 \\ 0.02 & 0.4 & 0.06 \\ -0.05 & 0.06 & 0.6 \end{pmatrix}, \tag{6.29}$$

The variance-covariance matrix is a symmetric matrix. It can be calculated with the cov function.

and the vector of expected returns is given by

$$\bar{\mathbf{R}} = \begin{pmatrix} 0.1 \\ 0.12 \\ 0.14 \end{pmatrix}. \tag{6.30}$$

6.3.1 Minimum Variance Portfolio

LET US CALCULATE THE WEIGHTS that will provide us with a portfolio with minimum risk. As mentioned above, the risk of the portfolio is related to the standard deviation and it can be calculated as follows[4]:

$$\sigma = \sqrt{\mathbf{X}^T \Sigma \mathbf{X}}, \tag{6.31}$$

[4] Elton, E., M. Gruber, S. Brown, and W. Goetzmann (2009). *Modern Portfolio Theory and Investment Analysis*. John Wiley & Sons

where $\mathbf{X}$ is a vector whose elements represent the weight on each of the assets in the portfolio. The expected return of the portfolio is given by

$$R_p = \mathbf{X}^T \bar{\mathbf{R}}. \tag{6.32}$$

The superscript ^{T} denotes the transpose operation. See Section 3.2.

We obtain the minimum variance portfolio by solving the following optimisation problem:

$$\min \mathbf{X}^T \Sigma \mathbf{X}$$
$$\text{subject to:} \tag{6.33}$$
$$\sum_i X_i = 1,$$

The weights X_i can be thought of as the percentages invested in each asset and thus they must add up to 100%.

where $\mathbf{X}^T \Sigma \mathbf{X}$ is the variance of the portfolio and we are using the constraint that the weights of the portfolio add up to 1.

We solve this problem using Lagrange multipliers to minimise the variance. In the case of the variance-covariance matrix shown in Equation (6.29) this leads us to the following linear system:

Lagrange multipliers is a popular optimisation technique finding local maxima or minima of a function subject to equality constraints.

$$\begin{pmatrix} 0.6 & 0.04 & -0.1 & -1 \\ 0.04 & 0.8 & 0.12 & -1 \\ -0.1 & 0.12 & 1.2 & -1 \\ 1 & 1 & 1 & 0 \end{pmatrix} \begin{pmatrix} x_1 \\ x_2 \\ x_3 \\ \lambda \end{pmatrix} = \begin{pmatrix} 0 \\ 0 \\ 0 \\ 1 \end{pmatrix}. \tag{6.34}$$

We can easily solve this linear system in MATLAB and Octave with the following script:

```
% Script Ch6_Portfolio.m
clear;
% Defining the variance-covariance matrix
sigma = [0.3 0.02 -0.05; 0.02  0.4  0.06; ...
      -0.05 0.06 0.6];


% Constructing the Lagrange multiplier matrices
a=ones(3,1);
A=[2*sigma a; a' 0]
b=[0; 0; 0; 1];
% Solving the linear system
x = A\b
```

We define the variance-covariance matrix.

The Lagrange multipliers matrix can be constructed with concatenation and the linear system solved with the left-division operator. See Section 3.10.

We now run the script to obtain the following output:

```
> Ch6_Portfolio

A =

    0.60000    0.04000   -0.10000    1.00000
    0.04000    0.80000    0.12000    1.00000
   -0.10000    0.12000    1.20000    1.00000
    1.00000    1.00000    1.00000    0.00000
x =

    0.47765
    0.28254
    0.23981
   -0.27391
```

We show the Lagrange multipliers matrix, and the solution to the linear system.

We have written the code such that the constructed matrix is displayed, so that we are able to verify that it corresponds to that in Equation (6.34). Finally, we can see that the weights of the portfolio are

$$X_1 = 0.47765, \tag{6.35}$$

$$X_2 = 0.28254, \tag{6.36}$$

$$X_3 = 0.23981, \tag{6.37}$$

These are the weights of the minimum variance portfolio.

and finally the Lagrange multiplier is $\lambda = -0.27391$. Let us check that the weights add up to 1:

```
> MVP_weights=x(1:3);
> sum(MVP_weights)


ans =
      1
```

The colon notation lets us select the weights (Section 3.7) and we then use the sum function to add the elements of the sub-vector (see Section 5.6.9).

We now implement two anonymous functions that enable us to calculate the risk and return for the minimum variance portfolio calculated above. For the risk we have the following:

```
> risk = @(sigma, weights) ...
      sqrt(weights'*sigma*weights);
> MVP_Risk=risk(sigma,  MVP_weights)


MVP_Risk =
      0.37008
```

This anonymous function takes the variance-covariance and weights arrays to calculate the risk. See Section 5.7.1 for information about anonymous functions.

For the expected return of the portfolio we have

```
> R_portfolio = @(R_bar, weights) weights'*R_bar;
> R_bar = [0.1; 0.12; 0.14];
> MVP_Return = R_portfolio(R_bar,MVP_weights)


MVP_Return =
      0.11524
```

In this case the inputs of the anonymous function are the returns and weights arrays to calculate the expected return.

So, the risk of the minimum variance portfolio is approximately 37.0% and the return of the portfolio is approximately 11.52%.

6.3.2   Target Portfolio

LET US NOW CONSIDER USING the same variance-covariance matrix and expected return shown in Equations (6.29) and (6.30) to find the efficient portfolio with target return $\mu = 0.1$ when short-selling is allowed. We can solve this problem with the Lagrange multipliers method, and in this case we end up with the following linear system:

$$\begin{pmatrix} 2\Sigma & -\bar{\mathbf{R}} & -\mathbf{1} \\ -\bar{\mathbf{R}}^T & 0 & 0 \\ \mathbf{1}^T & 0 & 0 \end{pmatrix} \begin{pmatrix} \mathbf{X} \\ \lambda_1 \\ \lambda_2 \end{pmatrix} = \begin{pmatrix} \mathbf{0} \\ \mu \\ 1 \end{pmatrix}, \qquad (6.38)$$

The Lagrange multipliers technique can easily accommodate the constraint for a target return μ.

where the variance-covariance matrix Σ is an $n \times n$ symmetric matrix, with n being the number of assets in the portfolio; the vector of returns $\bar{\mathbf{R}}$ and the weights $\mathbf{X}$ are $1 \times n$ vectors. Finally, $\mathbf{1}$ represents a $1 \times n$ vector whose elements are all one and $\mathbf{0}$ is an $n \times 1$ vector of zeros.

We have decided to represent our problem in this way so that we are able to construct a function that can take as inputs the variance-covariance Σ, the expected return $\bar{\mathbf{R}}$ and the target return μ. In that way, not only can we solve for the target return mentioned above, but also for any other value.

The implementation of a function will enable us to reuse this calculation for any 3-asset portfolio.

We shall call this function `TargetPortfolio.m` and the code is as follows:

```
% Code for function TargetPortfolio.m
function [weights] = ...
        TargetPortfolio(sigma, expectedR, mu)
% This function calculates the weights for an
% n-asset portfolio with variance-covariance
% matrix sigma, and expected
% returns expectedR and target return mu


% Building a Lagrange multipliers matrix
n=size(sigma,1);
a = ones(n,1);
A = [2*sigma -expectedR -a; ...
        expectedR' 0 0; ...
        a' 0 0];
b= [zeros(n,1); mu; 1];


% Solving the linear system:
weights = A\b;
end
```

The input of this function are the variance-covariance matrix, the expected returns and the target return. The function will output the weights. For information about functions see Section 5.5.2.

The size lets us check the number of assets in the portfolio (see Section 3.1).

The functions ones and zeros (see Section 3.3) let us construct the relevant matrices by concatenation (see Section 3.5).

We now call the function to obtain the weights for the target portfolio with return $\mu = 0.1$ (with sigma and R_bar as defined above):

```
> myWeights=TargetPortfolio(sigma,R_bar,0.1)

myWeights =
    0.89904
    0.20192
   -0.10096
  -18.61058
    2.41865
```

We enter the inputs for our function and call it to obtain the weights that let us construct a portfolio with return $\mu = 0.1$.

The weights are thus given by

$$X_1 = 0.89904, \tag{6.39}$$
$$X_2 = 0.20192, \tag{6.40}$$
$$X_3 = -0.10096, \tag{6.41}$$

These are the weights for a target portfolio with $\mu = 0.1$.

and finally the Lagrange multipliers are $\lambda_1 = -18.61058$ and $\lambda_2 = 2.41865$. We could check that the weights obtained indeed add up to one:

```
> Target_weights = myWeights(1:3);
> sum(Target_weights)


ans =
      1
```

As before, we can use the colon notation and the sum function to verify that the weights add up to 1.

We finally use the anonymous functions defined above to calculate the risk and return of this target portfolio:

```
> Target_risk = risk(sigma, Target_weights)


Target_risk =
     0.52801


> Target_return=R_portfolio(R_bar, Target_weights)


Target_return =
     0.10000
```

The anonymous functions risk and R_Portfolio can be re-used to calculate the risk and return for this portfolio.

We can thus verify that the weights calculated create a portfolio with return $\mu = 0.1$.

As we can see, the risk associated with this target portfolio is $\sigma = 53.8\%$ and we corroborate that the weights calculated indeed imply a return of $\mu = 10\%$.

6.4   Differential Equations: Predator-Prey Model

DIFFERENTIAL EQUATIONS FIND APPLICATIONS IN a large
number of areas, ranging from physics to biology. In this
case let us take a look at a predator-prey model[5] based
on the Lotka-Volterra equations for a prey population, say
rabbits, $R(t)$ and a predator population, say foxes, $F(t)$:

[5] Britton, N. (2003). *Essential Mathematical Biology*. Springer Undergraduate Mathematics Series. Springer London

$$\frac{dR}{dt} = rR - \alpha RF, \qquad (6.42)$$

$$\frac{dF}{dt} = -\delta F + \beta RF. \qquad (6.43)$$

The model above assumes that without foxes, i.e. when
$F(t) = 0$, the rabbit population grows exponentially at a rate
r:

$$\frac{dR}{dt} = rR; \qquad (6.44)$$

similarly, without rabbits ($R(t) = 0$) the fox population
decreases exponentially at a rate δ:

$$\frac{dF}{dt} = -\delta F. \qquad (6.45)$$

THE INTERACTION OF THE PREY and predator populations
is represented by the terms that contain RF, having on the
one hand a negative impact on the rabbits, and on the other
a positive impact on the foxes. We are interested in using
MATLAB and Octave to model the change in size of the
rabbit and fox populations, together, over time.

MATLAB and Octave are able to
solve initial value problems with
built-in functions such as ode45 in
MATLAB and lsode in Octave.

Furthermore, we shall assign the following values to the
different parameters in Equations (6.42) and (6.43): $r = 1.2$,
$\alpha = 2.5$, $\delta = -0.5$ and $\beta = 0.6$. Also we will need the initial

values for the rabbit and fox populations: $R(0) = 80$ and $F(0) = 100$.

MATLAB AND OCTAVE ARE ABLE to solve the system of ordinary differential equations (ODEs) shown above by using, for instance, a built-in implementation of the Runge-Kutta method. In MATLAB, we can use the function ode45. Please note that MATLAB has a number of other solvers and they are used in a very similar fashion to the one highlighted here.

ode45 is an implementation of the Runge-Kutta method of order 4.

The usage of the ode45 function is as follows:

```
[t,y] = ode45(@f_handle, [t0,tf], initial_vals)
```

MATLAB

where @f_handle is a handle to a function that defines the differential equations to be solved, [t0,tf] is a vector that defines the initial and final values of the time variable t and finally initial_vals is a vector with the initial conditions for the spatial vector y.

Octave does not have an implementation of ode45; instead you can use the lsode function. The inputs that lsode expects are similar to those passed to MATLAB's ode45. Let us take a look:

lsode can be used in Octave to solve stiff ordinary differential equations.

```
[y] = lsode(@f_handle, initial_vals, t_interval)
```

Octave

As we can see the variables are the same as defined above, and t_interval is a vector that defines the time interval.

Notice that the order of the input for ode45 and lsode is different and this is an important point, as this has an impact on the script that defines the function handle to be passed to lsode compared to that passed to ode45.

Octave's lsode requires the inputs in reverse order to MATLAB's ode45.

Before tackling the solution of the coupled system of differential equations, let us rewrite our model in a vectorised form: we define a vector **v** whose first component v_1 refers to the rabbit population whereas the second component v_2 refers to the fox one:

$$\frac{d\mathbf{v}}{dt} = \frac{d}{dt}\begin{bmatrix} v_1 \\ v_2 \end{bmatrix},$$

$$= \begin{bmatrix} \frac{dR}{dt} \\ \frac{dF}{dt} \end{bmatrix},$$

$$= \begin{bmatrix} 1.2R - 2.5RF \\ -0.5F + 0.6RF \end{bmatrix}. \qquad (6.46)$$

It is important to express the system of ordinary differential equations in terms of matrices. This will be needed to define a function for MATLAB and Octave to use an input.

We will use this vectorised form of our problem to write a function that implements the model given by Equations (6.42) and (6.43). The function that defines the ODE system in MATLAB is a bit different to the one in Octave.

6.4.1 Ordinary Differential Equation System in MATLAB

WE NEED TO DEFINE A function that implements the ODE system and which can be passed as a handle to ode45. You are encouraged to take a look at the help offered for this function by typing help ode45. In a nutshell, this function expects a definition where the input for the time variable t is given first, and then the space vector (v in our case).

Remember that MATLAB expects the independent variable t first and then the dependent variables in vector form.

With this in mind, let us create a function script which we will call predator_prey_matlab.m:

```
function dv = predator_prey_matlab(t,v)
% This function calculates
% the Lotka-Volterra equations
% for predator-prey modelling
% for use in MATLAB
%     dR/dt = 0.5R - 0.01 R*F
%     dF/dt = -0.5R + 0.01 R*F


% Using a vectorised version where
% v=[v(1) v(2)]=[R F]


%  Initialise the vector dv with zeros:
dv=zeros(2,1);


%  Calculate dR/dt
dv(1) = 0.5*v(1) - 0.01*v(1)*v(2);


%  Calculate dF/dt
dv(2) = -0.5*v(2) + 0.01*v(1)*v(2);


end
```

MATLAB

The function will return a vector whose entries are given by the ODEs in question. For function definitions see Section 5.5.2.

We first initialise the elements with zeros (see Section 3.3).

Finally, we implement each of the ODEs in the system.

In the function above we have defined a vector dv whose entries dv(1) and dv(2) hold the values for $\frac{dR(t)}{dt}$ and $\frac{dF(t)}{dt}$. We initialise the vector with zeros and assign to each component Equations (6.42) and (6.43), respectively.

6.4.2   Ordinary Differential Equation System in Octave

THE FUNCTION THAT WILL BE used for lsode requires that the space vector v is passed first, followed by the time variable t. As such, the only thing we need to change

Octave expects a function where the dependent variables are first and then the independent variable t.

from the code shown for `predator_prey_matlab.m` is the first line. For completeness, let us show the entire function `predator_prey_octave.m`:

```
function dv = predator_prey_octave(v,t)
% This function calculates
% the Lotka-Volterra equations
% for predator-prey modelling
%      dR/dt = 0.5R - 0.01 R*F
%      dF/dt = -0.5R + 0.01 R*F
% for use in Octave


% Using a vectorised version where
% v=[v(1) v(2)]=[R F]


%  Initialise the vector dv with zeros:
dv=zeros(2,1);


%  Calculate dR/dt
dv(1) = 0.5*v(1) - 0.01*v(1)*v(2);


%  Calculate dF/dt
dv(2) = -0.5*v(2) + 0.01*v(1)*v(2);


end
```

Octave

The function will return a vector whose entries are given by the ODEs in question. For function definitions see Section 5.5.2.

We first initialise the elements with zeros (see Section 3.3).

Finally, we implement each of the ODEs in the system.

6.4.3   Solving the Predator-Prey System

WE ARE NOW ABLE TO write a script, let us call it `SolvingPredatorPrey.m`, that is able to solve the differential equation system using either `ode45` for MATLAB or `lsode` for Octave.

Let us start by defining useful parameters for this problem as follows:

```
% Script SolvingPredatorPrey.m:
% Solve the system of ODEs defined in
% the function predator_prey...

% If mlab = 1 we are using MATLAB
% otherwise we are using Octave.
mlab = 1;

% Define initial and final times
t0 = 0;
tf = 50;
n=1000;

% Used in  Octave only
t_interval = linspace(t0,tf,n);

% Define initial values vector
% with R(0) = 80, and F(0) = 100
initial_vals = [80; 100];
```

To allow for flexibility in using MATLAB and Octave, we implement the control variable mlab to decide if we are running the code in MATLAB or Octave.

We define initial and final times for the solution.

Similarly, we define the initial values.

The script SolvingPredatorPrey.m is written in such a way that it can be used with either of the two ODE system definitions shown in the previous sections. We are able to do so with the use of a control variable; in this case we call this variable mlab: when mlab is equal to 1 we implement the solution using MATLAB's ode45; for any other value (0 for example) the solution will be found using Octave's lsode. We now proceed to solve the system:

```
% Solving the differential equations
if mlab == 1
    [t,v] = ode45(@predator_prey_matlab,[t0,tf],...
      initial_vals);
else
    [v] = lsode(@predator_prey_octave, ...
      initial_vals, t_interval);
     t=t_interval';
end


% Returning to original functions R(t) and F(t)
R = v(:,1);   F = v(:,2);


% Opening a file to store R(t) and F(t)
fid=fopen('PredatorPrey.txt','w');
% Printing the values to the file
fprintf(fid,'%12.8f %12.8f %12.8f\n', ...
    [t'; R'; F']);
% Closing the file
fclose(fid);
```

The control variable mlab lets us use ode45 with MATLAB or lsode with Octave. Remember that the inputs have different order.

Finally, we extract the populations and write the results to a file called PredatorPrey.txt with the aid of fopen and fprintf. See Section 5.10.

Once we have solved the ordinary differential equation system with either MATLAB's ode45 or Octave's lsode we output the solution to a file. In other words, when we run the script SolvingPredatorPrey.m we will obtain a text file called PredatorPrey.txt that contains the population for rabbit $R(t)$ and foxes $F(t)$ at each time t.

We can further write a script PlottingPredatorPrey.m that lets us read the content of the file and plot it. We can achieve this with the help of fopen and fscanf to read the contents of the file in question and use subplot to visualise the functions:

```
% Script PlottingPredatorPrey.m

% Reading the file for R(t) and F(t)
clear;
fid=fopen('PredatorPrey.txt','r');
A=fscanf(fid,'%12f %12f %12f',[3 inf]);
fclose(fid);

% Extracting parts of A with
% meaningful names
t=A(1,:)'; R=A(2,:)'; F=A(3,:)';

% Plot R(t) values in red with *,
% F(t) values in black with circles.

% First Panel
subplot(2,1,1);
plot(t,R,'r*-',t,F,'ko:')
axis([0 50 0 150]);
xlabel('Time, t'); ylabel('Population Sizes');
title('a)Model Predator Prey, Solutions ...
    Over Time')
legend('Rabbits','Foxes')

%Second Panel
subplot(2,1,2);
plot(R,F,'k')
xlabel('Rabbits, R(t)'); ylabel('Foxes, F(t)');
title('b) Phase Plane for Predator Prey Model')
```

This script lets us read the contents of the file PredatorPrey.txt.

We assign an id to the file and open it with fopen. Reading can be done with fscanf. See Section 5.10.

We can now visualise the solution to the ODE system with the help of subplot. See Section sec:subplots.

Running the script above results in the plots shown in Figure 6.1 where we can see, in panel a), the dynamics of the rabbit and fox populations over time, and in panel b) the behaviour in phase space.

a) Model Predator Prey, Solutions Over Time

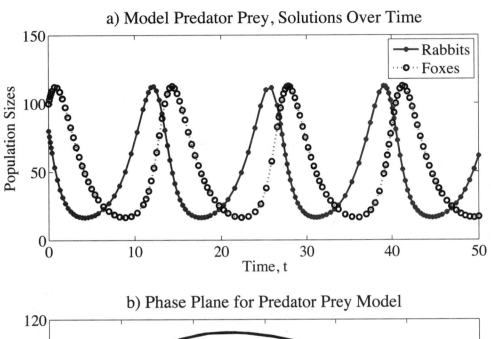

b) Phase Plane for Predator Prey Model

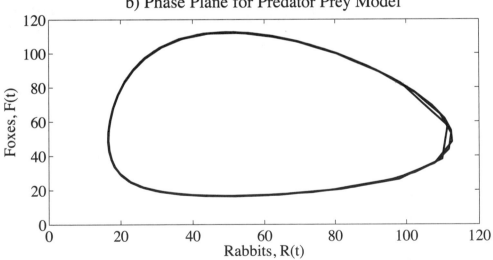

Figure 6.1: Solving the predator-prey model given by Equations (6.42) and (6.43). Panel a) shows the dynamics of the predator and prey populations, whereas panel b) shows the behaviour in phase space.

6.5   Signal Processing: Fourier Transform

SIGNALS THAT REPEAT THEMSELVES AFTER a certain period of time are called periodic. The most fundamental periodic signal is the sinusoidal one, and as such any other periodic signal can be thought of as the sum of sinusoidals with different amplitudes and frequencies[6]. One way to determine the components of this sum is by using the Fourier transform. Computationally, we shall make use of the fast Fourier transform algorithm (fft), which is implemented in MATLAB and Octave.

[6] Priemer, R. (1991). *Introductory Signal Processing*. Advanced Series in Electrical and Computer Engineering. World Scientific

Imagine that we have a signal given by the following mathematical function:

$$y = 2\sin(6\pi t) + 8\sin(9\pi t), \qquad (6.47)$$

and we are interested in sampling this signal during an interval of 2 seconds at a sampling rate of 0.01. This can be done with the script Sampling.m:

Some signal processing can be done with the aid of the Fourier transform. MATLAB and Octave have it implemented in the built-in function fft. For related functions see Section 5.6.15.

```
% Sampling.m

% Define sampling rate and interval
interval = 2;
dt = 1/100;
t = 0:dt:interval;
y= 2*sin(6*pi*t)+8*sin(9*pi*t);


% Plotting the signal
plot(t,y); grid on;
xlabel('Time (s)')
ylabel('Amplitude')
```

We can visualise the sampling with the plot command. See Section 4.1.

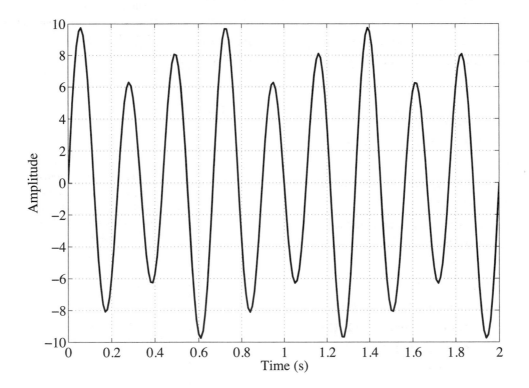

Figure 6.2: Sample of the signal given by Equation (6.47) for an interval of 2 seconds at a sample rate of 0.01.

The shape of the signal given by Equation (6.47) can be seen in Figure 6.2.

6.5.1 Amplitude Spectrum

GIVEN A SINUSOID SUCH AS

$$y(t) = A \sin(2\pi f t + \phi), \qquad (6.48)$$

we know that A is the amplitude, f is the ordinary frequency and ϕ is the phase. With this in mind, Equation (6.47) tells us that the frequencies that make up the signal

y are thus 3 Hz and 4.5 Hz, and that the amplitudes of the original signals are 2 and 8, respectively. We can identify these amplitudes and frequencies by calculating the amplitude spectrum of the signal using the Fourier transform:

The amplitude spectrum reveals the main frequencies that make up the signal.

```
% AmplitudeSpectrum.m
% Given the signal y calculate the FFT
Y=fft(y);
% Calculate the normalisation constant
n=size(y,2)/2;
% Take the abs value and normalise
spectrum=abs(Y)/n;

% Plotting the FFT
frequencies = (0:14)/(2*n*dt);
plot(frequencies, spectrum(1:15)); grid on
xlabel('Frequency (Hz)');
ylabel('Amplitude');
```

We take the Fourier transform of the signal with fft and a normalisation constant is calculated. For more information check the help for this function.

We then plot the amplitude spectrum with plot.

The plot generated with the script above can be seen in Figure 6.3 where we are able to clearly distinguish two peaks, one at 2 Hz and the other at 4 Hz as expected. Furthermore, the heights of the peaks indicate the amplitudes of the original sinusoidal signals.

6.5.2   Noise Filtering

THE FOURIER TRANSFORM USED IN the example above can be used to filter out noise from a signal. Let us consider adding some random noise to the signal y given by Equation (6.47):

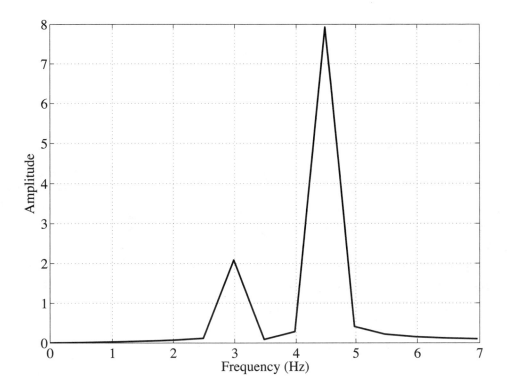

Figure 6.3: Amplitude spectrum of the signal given by Equation (6.47). Notice the peaks at the frequencies of 3 Hz and 4.5 Hz, with amplitudes 3 and 8, respectively.

```
% Script NoiseFiltering.m

% Generating random noise
noise = randn(1,size(y,2));
noisy_y = y + noise;

% Amplitude spectrum of noisy signal
noisy_Y = fft(noisy_y);
n = size(noisy_y,2)/2;
noisy_spectrum = abs(noisy_Y)/n;
```

We first add some random noise to our signal with the help of randn. See Section 5.2.4 for an example. We then calculate the amplitude spectrum as above.

We can add to our script a few commands to visualise the
noisy signal and its amplitude spectrum:

```
% Plotting noisy signal
figure(1)
subplot(2,1,1);
plot(time,noisy_y); grid on
title('a) Signal with random noise');
xlabel('Time (s)');
ylabel('Amplitude');


% Plotting amplitude spectrum
subplot(2,1,2);
noisy_freq = (0:14)/(2*n*dt);
plot(noisy_freq,noisy_spectrum(1:15)); grid on
title('b) Amplitude spectrum of noisy signal');
xlabel('Frequency (Hz)');
ylabel('Amplitude');
```

The command subplot enables
to visualise both the noisy signal
and its frequency spectrum. See
Section 4.5.

The result of adding noise to our signal can be seen in panel
a) of Figure 6.4. We can compare this to the signal shown in
Figure 6.2 and observe that they share some characteristics.
In order to convince ourselves that the main frequencies
are the same, we plot the amplitude spectrum of the noisy
signal in panel b) of Figure 6.4 where we can see the same
peaks that we have observed before.

The amplitude spectrum of the
noisy signal shows its main frequencies enabling us to implement
a filter.

Finally, let us use the inverse Fourier transform to filter out
the low amplitude noise to correct the signal. We do this
by eliminating those points that are above a cut off value in
amplitude:

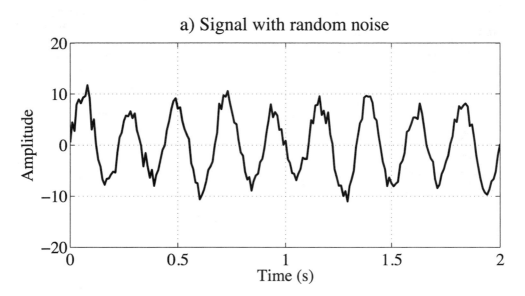

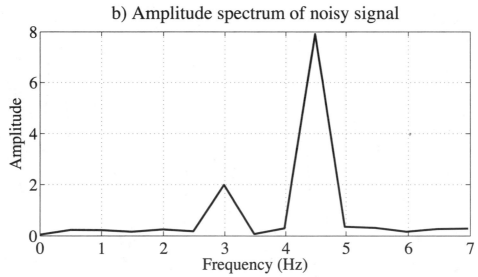

Figure 6.4: Panel a) Noisy signal generated by adding random noise to original sampling. Panel b) Amplitude spectrum of the noisy signal.

```
% Cancelling noise
fixed_Y = noisy_Y.*(abs(noisy_Y)>100);
% Inverting the FFT and keeping the real part
inv_fixed_Y = ifft(fixed_Y);
corrected_y = real(inv_fixed_Y);
% Plotting the corrected signal
figure(2)
plot(time,corrected_y); grid on
title('Corrected signal')
xlabel('Time (s)');
ylabel('Amplitude');
```

High amplitudes are eliminated by keeping terms below a certain threshold. See Section 5.2.4. We then apply the inverse Fourier transform, ifft, to recover the filtered signal.

We than visualise the filtered signal with a plot.

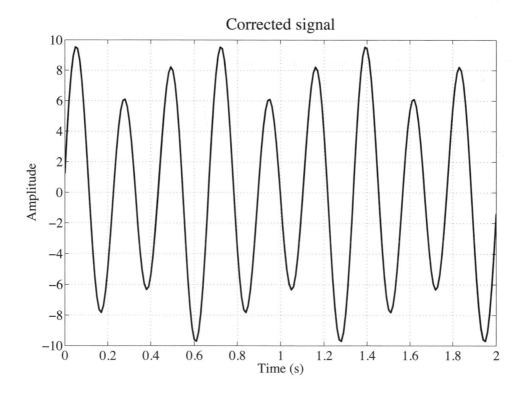

Figure 6.5: Corrected signal after applying a low amplitude filter to the noisy input.

As we can see from Figure 6.5 the filtering has worked quite well and thus we have recovered a fairly good match for our original signal.

6.6   Physics: The Wave Equation

THE WAVE EQUATION PROVIDES A straightforward way to model the propagation of waves of different kinds: light, sound, oscillating strings and membranes, etc. The wave equation in one dimension is given by the following partial differential equation[7]:

$$\frac{\partial^2 E}{\partial t^2} - c^2 \frac{\partial^2 E}{\partial x^2} = 0, \tag{6.49}$$

[7] Elmore, W. and M. Heald (2012). *Physics of Waves*. Dover Books on Physics. Dover Publications

where $E = E(x, t)$ is a function that gives the amplitude of the wave at time t in position x, and c is the speed with which the wave propagates.

6.6.1   Oscillations in a String

WE ARE INTERESTED IN MODELLING the oscillations of a string, for instance in a guitar. Let us consider that the length of the string is 10 cm and that the wave propagates at a speed of 5 cm/s. Both ends of the string are fixed at a height $x = 0$.

We model the oscillations in a guitar string for example.

We tackle the solution of the wave Equation (6.49) using finite differences. Let us recall that the second derivative of a function can be approximated with the use of central differences as:

$$f''(x) \simeq \frac{1}{\Delta x^2} \left(f(x + \Delta x) - 2f(x) + f(x - \Delta x) \right). \tag{6.50}$$

Finite difference approximation for the second derivative.

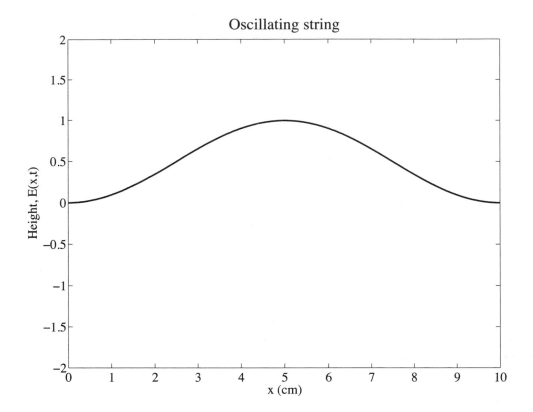

Figure 6.6: Initial condition for the simulation of an oscillating string.

This means that we can express Equation (6.49) as

$$\frac{1}{\Delta t^2}\left(E(x,t+\Delta t) - 2E(x,t) + f(x,t-\Delta t)\right) =$$

$$\frac{c^2}{\Delta x^2}\left(E(x+\Delta x,t) - 2E(x,t) + E(x-\Delta x,t)\right). \qquad (6.51)$$

This expression depends on earlier, current and future times.

Notice that we have three different times in play here: the time now (t), later times ($t + \Delta t$) and earlier times ($t - \Delta t$). We can use this to our advantage such that we can predict the position at later times based on current and past positions.

We use the expression above to solve for later times, i.e. $t + \Delta t$:

$$\begin{aligned} E(x, t + \Delta t) &= r^2 \left(E(x + \Delta x, t) + E(x - \Delta x, t) \right) + \\ &\quad 2(1 - r^2) E(x, t) - E(x, t - \Delta t), \quad (6.52) \end{aligned}$$

We have expressed the position at later times in terms of current and past positions.

where $r = \frac{c \Delta t}{\Delta x}$. If we choose Δt to be small enough such that $r < \frac{1}{2}$, then the numerical solution will be stable; otherwise we end up with ever larger waves. We shall consider the initial condition

$$E(x, 0) = \frac{1}{2} \left(1 - \cos \left(\frac{2 \pi x}{L} \right) \right), \quad (6.53)$$

We need to provide an initial condition for the simulation.

as shown in Figure 6.6.

The script `StringOscillation.m` implements the simulation. We shall divide the script into two parts. First, let us set up the parameters of the simulation as follows:

```
% Script StringOscillation.m
% Defining parameters
delta_x = 0.01;
delta_t = 0.001;
c = 5;
L = 10;
end_sim=30;

r=c*delta_t/delta_x;
n = L/delta_x +1;
string = [0:delta_x:L];

% Setting initial condition to a cos function
now_t = (1/2)*(1-cos(2*pi/L*string));
earlier_t=now_t;
```

We define the values for Δx and Δt, as well as the values for parameters such as L, c and the duration of the simulation.

This enables us to define the discretisation of the length of the string and set up the initial condition.

We implement the finite difference update outlined in
Equation (6.52). The result of the script below results in an
animated plot, and thus we recommend trying the script to
see the end result.

This script will result in an
animation.

```
% Wave equation solution
for time=0:delta_t:end_sim
  % calculate future position
  later_t(1)=0;
  later_t(2:n-1)=r^2*(now_t(1:n-2)+...
     now_t(3:n)) + ...
     2*(1-r^2)*now_t(2:n-1) - earlier_t(2:n-1);
  later_t(n) = 0;

  % update status for next iteration
  earlier_t = now_t;
  now_t = later_t;

  % plot the status every 20th frame
  if mod(time/delta_t, 20)==0
    plot(string,now_t);
    title('Oscillating string');
    xlabel('x (cm)');
    ylabel('Height, E(x,t)');
    axis([0 L -2 2])
    pause(0.001)
  end
end
```

We implement the finite difference
scheme, fixing the end points of
the string.

On each iteration we update the
time (see Section 5.3). Finally, we
generate a plot.

6.6.2   Oscillations in a Circular Membrane

IN THE PREVIOUS SECTION WE have seen how the oscillations in a string can be modelled with the aid of MATLAB

and Octave. Here, we shall tackle a related problem, but instead of a string in a guitar, we are interested in modelling the oscillations in a drum. We can think of this as an elastic 2D membrane stretched over a circular frame of radius a. In this case the wave equation is best treated in polar coordinates[8]:

[8] Rao, S. (2007). *Vibration of Continuous Systems*. Wiley

$$\frac{\partial^2 E}{\partial t^2} = c^2 \left(\frac{\partial^2 E}{\partial r^2} + \frac{1}{r}\frac{\partial E}{\partial r} + \frac{1}{r^2}\frac{\partial^2 E}{\partial \theta^2} \right), \qquad (6.54)$$

We are now dealing with a system in two dimensions in a circular domain, akin to a drum. We thus use polar coordinates to tackle this problem.

where $E = E(r,\theta,t)$ is a function of the radial and polar variables r and θ and time t, with $0 \le r \le a$ and $-\pi \le \theta \le \pi$.

Using separation of variables such that $E(r,\theta,t) = R(r)\Theta(\theta)T(t)$ it is possible to show that, up to scaling, rotation and phase shift in time, the solution to the vibrations of a 2D circular membrane is given by

$$E(r,\theta,t) = J_m\left(\lambda_{mn}r\right)\cos\left(m\theta\right)\cos\left(c\lambda_{mn}\right), \qquad (6.55)$$

for $m = 0,1,2,\ldots$, $n = 1,2,3,\ldots$ and where J_m are the Bessel functions of order m of the first kind, with $\lambda_{mn} = \frac{\alpha_{mn}}{a}$ and α_{mn} being the n-th positive zero of J_m. Since the membrane is attached to the circular frame of the drum, we have a boundary condition that requires us to have a node at $r = a$.

J_m is the Bessel function of order m. See Section 5.6.6.

The solution above means that we require a pair of indices (m,n) to specify in a unique way the vibrational modes of the circular membrane; furthermore, we also require to know the zeros of the Bessel functions to characterise the solution of the wave Equation (6.54). Let us use this information to write the script drum.m that simulates various modes of vibration for this 2D circular membrane. We first set up the parameters for the simulation:

The implementation of the solution requires two indices (m,n) as well as knowing the zeros of the Bessel function.

```
% Script drum.m - 2D membrane oscillations
clear
% Setting up parameters
m=4;
n=2;
c=0.1;

% Defining r and theta
r=linspace(0,1,51);
theta=linspace(-pi,pi,201);
[r,theta]=meshgrid(r,theta);
% Transforming polar to cartesian
[x,y]=pol2cart(theta,r);


% Bessel function zeros
J0zeros = [2.40482555769577 5.52007811028631 ...
    8.65372791291101 11.7915344390142 ...
    14.9309177084877];
J1zeros = [3.83170597020751 7.01558666981561 ...
    10.1734681350627 13.3236919363142 ...
    16.4706300508776];
J2zeros = [5.13562230184068 8.41724414039986 ...
    11.6198411721490 14.7959517823512 ...
    17.9598194949878];
J3zeros = [6.38016189592398 9.76102312998166 ...
    13.0152007216984 16.2234661603187 ...
    19.409415226435];
J4zeros = [7.58834243450380 11.0647094885011 ...
    14.3725366716175 17.6159660498048 ...
    20.8269329569623];
Js=[J0zeros; J1zeros; J2zeros; J3zeros; J4zeros];
```

These parameters can be modified to generate various modes.

The coordinates r and θ can be easily defined. We then use the function pol2cart to map the values to rectangular coordinates (see Section 5.6.8).

In this case we provide a list of the zeros of the first five Bessel functions J_m, however MATLAB and Octave can be used to find them.

With this in place we tackle the simulation as follows:

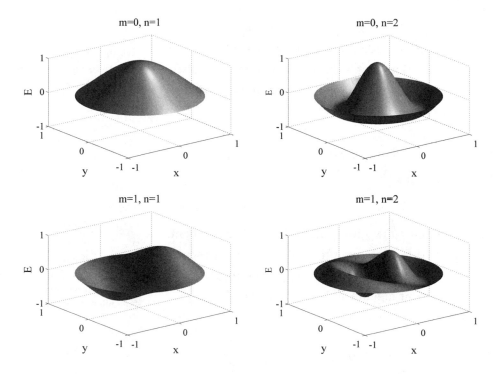

Figure 6.7: Modes of a 2D circular membrane, for m=0,1 and n=1,2.

```
% Simulation
time = linspace(0,10,50);
for t=time
  lambda=Js(m+1,n);
  u=real(besselj(m,r*(lambda))).*cos(m*theta) ...
    .*cos(c*lambda*t);
  mesh(x,y,u);
  xlabel('x'); ylabel('y'); zlabel('E');
  title(sprintf('m=%d, n=%d', m , n));
  axis([-1 1 -1 1 -1 1]);
  pause(0.1);
end
```

For the simulation we calculate the corresponding value for $\lambda_{m,n}$; remember that our drum has radius $a = 1$. We then implement the solution and visualise it in a 3D plot with the aid of mesh. See Section 4.8.

The result of the simulation above is a series of frames in the same figure, providing a rudimentary animation. We recommend trying the code above in your own computer to see the modes in action. To give us an idea of what we shall see, we have plotted some of the vibrational modes for the 2D circular membrane in Figure 6.7 .

6.7   Quantum Mechanics: The Schrödinger Equation and Pauli Matrices

THE USE OF MATRICES IN quantum mechanics is prominent: they are employed to represent physical quantities, describing states on a quantum system. This approach requires the constant use of linear algebra and what better tools than MATLAB and Octave to do so.

Linear algebra lies at the heart of quantum mechanics. Using MATLAB and Octave can facilitate some relevant operations.

6.7.1   Particle in an Infinite Potential Well

ONE OF THE SIMPLEST SYSTEMS in quantum mechanics is a particle confined in an infinite potential well. In this case we are interested in describing the discrete energy levels using the time-independent Schrödinger equation[9].

[9] Griffiths, D. (2005). *Introduction to Quantum Mechanics*. Pearson Education

Let us consider a quantum mechanical particle with wave-function $\psi(x)$ in one dimension. The particle is confined to a square well with impenetrable walls in the region $0 < x < L$. This means that we have boundary conditions that require the wavefunction to vanish at both ends, i.e. $x = 0$ and $x = L$. We therefore expect to end up with stable standing waves inside the box.

The wavefunctions for the states in a one-dimensional box are standing waves.

The time-independent Schödinger equation is given by

$$\hat{H}\psi(x) = E\psi(x),\tag{6.56}$$

The Schrödinger equation is an eigenvalue problem.

where E is the energy and $\hat{H}$ is the Hamiltonian associated with the system, and in this case it is given by

$$\hat{H} = -\frac{\hbar^2}{2m}\frac{\partial^2}{\partial x^2} + V(x), \qquad (6.57)$$

where $V(x)$ is the potential given as

$$V(x) = \begin{cases} 0, & 0 < x < L, \\ \infty, & \text{elsewhere.} \end{cases} \qquad (6.58)$$

The Hamiltonian is an operator that corresponds to the total energy in a system.

Equation (6.56) is effectively an eigenvalue problem and we therefore need to find the eigenvalues E and their associated eigenvectors to describe the quantised energy levels of the system.

We can tackle this problem with the use of the eig function. See Section 6.2.

With the aid of MATLAB and Octave we can solve this problem as follows: let us start by defining the parameters to be used in this problem.

```
% InfiniteWell.m - Schroedinger Equation
clear;

% Defining parameters
% Working in natural units
hbar=1;
m=1;
L = 2*pi;

% Number of points
N=100;
delta_x = L/(N-1);
x = 0:delta_x:L;
```

We define parameters that define the problem. Notice that we are working in natural units ($\hbar = 1$).

We also define a discretisation of the length of the box.

We are working in natural units where $\hbar$ has the value of 1, and the mass of the particle is also 1. The box has a length

$L = 2\pi$ and we have discretised our space variable x in 100 equidistant points.

In order to tackle this problem we require a way to calculate the first and second derivatives. This can be done by constructing differentiation matrices with 2 and 3 point-schemes:

Differentiation matrices are arrays whose entries are the weights of finite difference schemes similar to those used in Section 6.6.

```
% First derivative: 2pt finite-difference matrix
% Need f(0)=f(L)=0
Deriv=(diag(ones((N-1),1),1)- ...
   diag(ones((N-1),1),-1))/(2*delta_x);
% Defining the upper and lower ends ...
% of the matrix
Deriv(1,1) = 0;
Deriv(1,2) = 0;
Deriv(2,1) = 0;
Deriv(N,N-1) = 0;
Deriv(N-1,N) = 0;
Deriv(N,N) = 0;
% Laplacian: 3pt finite-difference matrix
Laplacian = (-2*diag(ones(N,1),0) + ...
   diag(ones((N-1),1),1) + ...
   diag(ones((N-1),1),-1))/(delta_x^2);
% Defining the upper and lower ends ...
% of the matrix
Laplacian(1,1) = 0;
Laplacian(1,2) = 0;
Laplacian(2,1) = 0;
Laplacian(N,N-1) = 0;
Laplacian(N-1,N) = 0;
Laplacian(N,N) = 0;
```

For the first derivative we use a central difference scheme with two points.

For the Laplacian, we implement a second derivative approximation using central differences with three points.

We now construct the Hamiltonian (Equation (6.57)) and find its eigenvalues:

```
% Constructing the Hamiltonian
H=-(hbar^2/(2*m))*Laplacian;


% Solving the eigenvalue problem
[Psi, E] = eig(H);


% Plotting the first 4 eigenfunctions
plot(x,Psi(:,3),'k-',x,Psi(:,4),'k:o', ...
   x,Psi(:,5),'k--+',x,Psi(:,6),'k-.');
legend(['n=1'; 'n=2'; 'n=3'; 'n=4'])
axis([0 L -0.2 0.2])
title('Eigenfunctions for a particle in a box')
xlabel('x')
ylabel('\psi(x)')
```

The Hamiltonian makes use of the Laplacian operation calculated above.

Solving the Schrödinger equation is equivalent to finding the eigenvalues of the Hamiltonian operator.

We can visualise the energy levels of a particle in a box with the aid of a plot. See Section 4.1.

The first four eigenfunctions for the quantum particle confined in an infinite well can be seen in Figure 6.8.

6.7.2   Pauli Spin Matrices

THE PAULI MATRICES ARE COMPLEX arrays that arise in the treatment of spin in quantum mechanics. In a unitless form they are given by[10]

[10] Griffiths, D. (2005). *Introduction to Quantum Mechanics*. Pearson Education

$$\sigma_x = \begin{pmatrix} 0 & 1 \\ 1 & 0 \end{pmatrix}, \tag{6.59}$$

$$\sigma_y = \begin{pmatrix} 0 & -i \\ i & 0 \end{pmatrix}, \tag{6.60}$$

$$\sigma_z = \begin{pmatrix} 1 & 0 \\ 0 & -1 \end{pmatrix}. \tag{6.61}$$

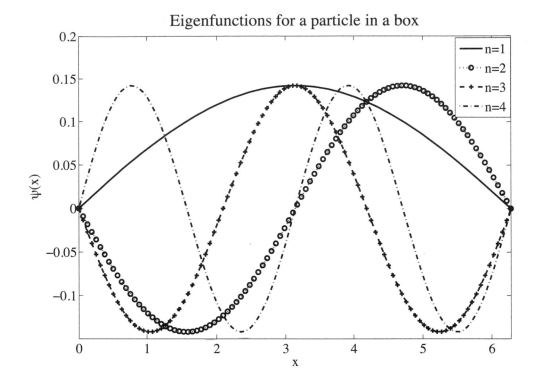

Figure 6.8: Eigenfunctions
n=1,2,3,4 for a quantum particle confined in an infinite potential
well.

Let us use MATLAB and Octave to show that the following
commutation relations hold:

$$[\sigma_x, \sigma_y] = 2i\sigma_z, \tag{6.62}$$

$$[\sigma_y, \sigma_z] = 2i\sigma_x, \tag{6.63}$$

$$[\sigma_z, \sigma_x] = 2i\sigma_y. \tag{6.64}$$

Let us recall that a commutator is an operation that tells us
if a binary operation, such as matrix multiplication, is commutative or not. Given two matrices **A** and **B** the commuta-
tor is calculated as $[\mathbf{A}, \mathbf{B}] = \mathbf{AB} - \mathbf{BA}$. The commutation of

We can check if the product of two
operators commute or not with
the aid of the commutator.

two operators in quantum mechanics is important because we cannot know the value of two physical quantities at the same time if the operators that represent them do not commute. This is related to the famous uncertainty principle postulated by Heisenberg.

We construct a function to calculate the commutator as:

```
function C = commutator(A, B)
% This function calculates the
% commutator [A, B]
    C=A*B - B*A;
end
```

A function to calculate the commutator of two operators.

We are now in a position to tackle the problem at hand; we start then by defining the Pauli matrices in MATLAB and Octave as follows:

```
> clear;
> sigma_x = [0 1; 1 0];
> sigma_y = [0 -1i; 1i 0];
> sigma_z = [1 0; 0 -1];
```

The Pauli matrices in MATLAB and Octave.

Remember that the imaginary number $i = \sqrt{-1}$ is denoted in MATLAB and Octave with the letters i and j. However, in order to avoid confusion, both environments also use 1i and that is the notation we have used above.

The complex number is denoted by i, j or 1i. See Section 1.5.1.

We can finally check that the commutation relations for the Pauli matrices hold. In this case we are going to show that

$$\sigma_z = \frac{[\sigma_x, \sigma_y]}{2i}, \tag{6.65}$$

$$\sigma_x = \frac{[\sigma_y, \sigma_z]}{2i}, \tag{6.66}$$

$$\sigma_y = \frac{[\sigma_z, \sigma_x]}{2i}. \tag{6.67}$$

```
> R1 = commutator(sigma_x,sigma_y)/(2*1i)


R1 =

      1      0
      0     -1


> R2 = commutator(sigma_y,sigma_z)/(2*1i)


R2 =

      0      1
      1      0


> R3 = commutator(sigma_z,sigma_x)/(2*1i)


R3 =
    0.0000 + 0.0000i    0.0000 - 1.0000i
    0.0000 + 1.0000i    0.0000 + 0.0000i
```

The result of the commutator divided by $2i$ indeed returns the third Pauli matrix.

We can see that by dividing the result of each commutator between two Pauli matrices (in correct cyclical order) by $2i$ we recover the third one.

6.8   Summary

As we have seen in this chapter, the flexibility and versatility offered by MATLAB and Octave make them suitable to tackle a wide range of problems and subjects.

In this case we have chosen a small number of topics with a few examples. Our main objective here has been to show how the different elements of the software are combined together to address a particular set of tasks within the context of particular applications. An in-depth discussion of these topics goes beyond the scope of this book.

We have seen how MATLAB and Octave can successfully be used for calculations directly involving linear algebra; matrices are, after all, the main element of the software. Optimisation problems are typical of a number of applications and in this case we have chosen to use portfolio management to exemplify the implementation of a very basic optimisation technique: Lagrange multipliers. Similarly, another important tool used in various areas is the numerical solution of differential equations; here we have used a predator-prey model to demonstrate the way to solve a system of ordinary differential equations. In the case of partial differential equations, we have used the wave equation as a way to show an implementation of finite differences to tackle this sort of problem.

Other applications included here are signal processing and noise filtering, the simulation of the oscillations in a circular membrane, as well as the solution of the time-independent Schrödinger equation for an infinite potential well and simple manipulations of spin matrices. We very much hope that these examples enable you, the reader, to use MATLAB and Octave in your own problem-solving tasks.

6.9   Exercises

The following exercises are related to the subjects discussed in this chapter. In order to tackle the exercises you are expected to be familiar with the topics and we recommend taking a look at the references used in this chapter for further details.

1. The angle θ between two vectors $\mathbf{u}$ and $\mathbf{v}$ can be obtained by recalling that the dot product of the two vectors can be expressed as

    $$\mathbf{u} \cdot \mathbf{v} = |\mathbf{u}||\mathbf{v}| \cos \theta.$$

    Write a function that takes any two vectors, checks if they have the same length and calculates the angle θ between them.

2. Write a function that takes a vector $\mathbf{v}$, and returns a unit vector $\hat{\mathbf{v}}$. Your functions should check for zero-norm vectors.

3. Find out what the functions `rank` and `rref` do. Write a function that takes a general linear system of linear equations $\mathbf{Ax} = \mathbf{b}$ with m equations and n variables. Determine whether the system has a unique solution and find the answer. Otherwise if the rank of the matrix $\mathbf{A}$ is different from the number of variables, display the augmented matrix using the `rref` function.

4. The Gram-Schmidt process is a procedure that builds an orthonormal basis over an arbitrary interval out of a given nonorthogonal set of linearly independent functions[11]. With the aid of an appropriate reference such as the one suggested, write an m-function that implements the Gram-Schmidt algorithm for n independent vectors $\mathbf{a_1}, \mathbf{a_2}, \ldots, \mathbf{a_n}$ to produce n orthonormal vectors $\mathbf{q_1}, \mathbf{q_2}, \ldots, \mathbf{q_n}$.

[11] Strang, G. (2003). *Introduction to Linear Algebra*. Wellesley-Cambridge Press

5. Consider the two three-asset portfolios used in Section 6.3 such that the weights of the minimum variance portfolio are given by the vector $\mathbf{w}_a$. Similarly, $\mathbf{w}_b$ are the weights of the portfolio with target return $\mu_b = 0.1$. For a situation where short-selling is allowed, write a script that calculates and plots the efficient frontier using a convex combination of portfolios $\mathbf{w}_a$ and $\mathbf{w}_b$.

6. Use MATLAB and Octave to solve and plot the solution to the following second-order differential equation:

$$y''(c) + 8y'(x) + 2y(x) = \cos(x); \qquad y(0) = 0, \quad y'(0) = 1.$$

Hint: Rewrite the equation as a first-order system.

7. Consider the Lorenz equations:

$$\frac{dx}{dt} = -55(x+y),$$
$$\frac{dy}{dt} = -y - 5xz,$$
$$\frac{dz}{dt} = xy - 4z - 420,$$

with the following initial conditions: $x(0) = -5$, $y(0) = 5$ and $z(0) = 15$. Solve the system of differential equations using MATLAB and Octave for a time interval $[0,8]$. Finally, produce a plot of $z(t)$ versus $x(t)$.

8. Consider the following signal:

$$y = 2\sin(6\pi t) + 8\sin(9\pi t).$$

Add some noise to it in the form of a sine wave of amplitude 5 and frequency 60 Hz. Plot the original and the noisy signals, as well as their amplitude spectra. Finally, use the Fourier transform to remove the high frequency noise and compare the results to the original signal.

9. Consider the diffusion equation:

$$\frac{\partial U}{\partial t} = \alpha \frac{\partial^2 U}{\partial x^2},$$

which can be used to model the temperature diffusion in a slab of length $L = 1$ with diffusivity $\alpha = 1 \times 10^{-2}$. Write a script to model the temperature diffusion in the slab for an initial condition given by

$$U(x,0) = \sin\left(\frac{2\pi x}{L}\right),$$

and subject to the boundary conditions $U(0,t) = 0$ and $U(L,t) = 0$.

10. Using the Pauli spin matrices we can represent the spin state of a spin-$\frac{1}{2}$ particle by a two-element column vector. For spin up and down these vectors are, respectively, given by

$$|\uparrow\rangle = \begin{pmatrix} 1 \\ 0 \end{pmatrix}, \qquad |\downarrow\rangle = \begin{pmatrix} 0 \\ 1 \end{pmatrix}.$$

(a) Verify that operating on these vectors from the left with the matrix σ_z yields $+|\uparrow\rangle$ and $-|\downarrow\rangle$, respectively.

(b) Construct a column vector $|\rightarrow\rangle$ with the property that $\sigma_x |\rightarrow\rangle = +|\rightarrow\rangle$, which corresponds to a spin-$\frac{1}{2}$ particle with spin in the $+\hat{x}$ direction.

(c) Finally, construct a column vector $|\otimes\rangle$ with the property that $\sigma_y |\otimes\rangle = +|\otimes\rangle$, which corresponds to a spin-$\frac{1}{2}$ particle with spin in the $+\hat{y}$ direction.

Differences between MATLAB® and Octave

MATLAB and Octave share quite a lot of functionality and it is possible to develop code in one of the programming environments to be successfully used in the other. However, it is inevitable that some differences manifest themselves from time to time. Here we present some of the main differences between the languages used by MATLAB and Octave. We would like to note that incompatibilities between both languages are addressed with every new version and thus we recommend checking the release notes on a regular basis.

- Octave supports the use of single and double quotes to define strings. MATLAB only supports single quotes.

- Octave supports C-style assignment and increment operators:

```
> i++;
> ++i;
> i+=1; % etc
```
Octave

- MATLAB requires ellipsis (...) for continuing lines:

```
> x = [1 2 3; ...
    4, 5, 6]
```
MATLAB

- Octave supports the use of ellipsis as above, plus it also allows for the following two forms:

```
> x = [1 2 3;
   4, 5, 6]


> x = [1 2 3;     4, 5, 6]
```
Octave

- In Octave we can specify data labels directly in the plot function, while MATLAB requires the use of the `legend` function.

```
> plot(x, y, ';MyLabel;')
```
Octave

- For ending procedure statements MATLAB requires end. In Octave, code blocks such as `for` and `while` loops as well as `if` statements can be terminated with `endfor`, `endwhile` and `endif`, respectively. The use of differentiated end commands can be very useful when using nested control structures in long scripts, as they may help in determining when a particular structure finishes.

- MATLAB uses the percentage symbol (%) to start a comment. Octave uses both the percentage symbol and the hash symbol (#).

- Octave supports C-style hexadecimal notation (e.g., "oxFo") whereas MATLAB requires the use of the `hex2dec` function:

```
> hex2dec('F0')
```
MATLAB

- MATLAB requires ^ for exponentiation. Octave can use ^ or **.

• The debugger that comes embedded in MATLAB has more advanced features such as graphical stops and continuation commands.

• MATLAB has a number of toolboxes that can be acquired separately from The MathWorks. Some implementations of similar capabilities are available for Octave. However some may not be as complete or may not be available.

• MATLAB is capable of producing Graphical User Interfaces (GUIs) with the help of guide. This capability is not available in Octave.

Bibliography

Britton, N. (2003). *Essential Mathematical Biology*. Springer Undergraduate Mathematics Series. Springer London.

Eaton, J. W., D. Bateman, and S. Hauberg (2008). *GNU Octave Manual: Version 3*. A GNU manual. Network Theory Limited.

Elmore, W. and M. Heald (2012). *Physics of Waves*. Dover Books on Physics. Dover Publications.

Elton, E., M. Gruber, S. Brown, and W. Goetzmann (2009). *Modern Portfolio Theory and Investment Analysis*. John Wiley & Sons.

GNU (June 29, 2007). General Public License, Free Software Foundation, version 3. http://www.gnu.org/licenses/gpl (Last visited Aug 4,2014).

Griffiths, D. (2005). *Introduction to Quantum Mechanics*. Pearson Education.

Hansen, J. S. (2011). *GNU Octave Beginner's Guide*. Learn by doing: less theory, more results. Packt Publishing, Limited.

Higham, D. J. and N. J. Higham (2005). *MATLAB Guide*. Society for Industrial and Applied Mathematics.

Palm, W. J. (2008). *A Concise Introduction to MATLAB*. McGraw-Hill Higher Education.

Priemer, R. (1991). *Introductory Signal Processing*. Advanced Series in Electrical and Computer Engineering. World Scientific.

Rao, S. (2007). *Vibration of Continuous Systems*. Wiley.

Strang, G. (2003). *Introduction to Linear Algebra*. Wellesley-Cambridge Press.

Index